Lecture Notes in Physics

Edited by J. Ehlers, München, K. Hepp, Zürich,
R. Kippenhahn, München, H. A. Weidenmüller, Heidelberg,
and J. Zittartz, Köln

63

V. K. Dobrev · G. Mack · V. B. Petkova
S. G. Petrova · I. T. Todorov

Harmonic Analysis

on the n-Dimensional Lorentz Group and Its Application to Conformal Quantum Field Theory

Springer-Verlag
Berlin · Heidelberg · New York

Lecture Notes in Physics

Bisher erschienen/Already published

Vol. 1: J. C. Erdmann. Wärmeleitung in Kristallen, theoretische Grundlagen und fortgeschrittene experimentelle Methoden. II, 283 Seiten. 1969.

Vol. 2: K. Hepp, Théorie de la renormalisation. III, 215 pages. 1969.

Vol. 3: A. Martin, Scattering Theory: Unitarity, Analyticity and Crossing. IV, 125 pages. 1969.

Vol. 4: G. Ludwig, Deutung des Begriffs „physikalische Theorie" und axiomatische Grundlegung der Hilbertraumstruktur der Quantenmechanik durch Hauptsätze des Messens. 1970. Vergriffen.

Vol. 5: Schaaf, The Reduction of the Product of Two Irreducible Unitary Representations of the Proper Orthochronous Quantummechanical Poincare Group. IV, 120 pages. 1970.

Vol. 6: Group Representations in Mathematics and Physics. Edited by V. Bargmann. V, 340 pages. 1970.

Vol. 7: R. Balescu, J. L. Lebowitz, I. Prigogine, P. Résibois, Z. W. Salsburg, Lectures in Statistical Physics. V, 181 pages. 1971.

Vol. 8: Proceedings of the Second International Conference on Numerical Methods in Fluid Dynamics. Edited by M. Holt. 1971. Out of print.

Vol. 9: D. W. Robinson, The Thermodynamic Pressure in Quantum Statistical Mechanics. V, 115 pages. 1971.

Vol. 10: J. M. Stewart, Non-Equilibrium-Relativistic Kinetic Theory. III, 113 pages. 1971.

Vol. 11: O. Steinmann, Pertubation Expansions in Axiomatic Field Theory. III, 126 pages. 1976.

Vol. 12: Statistical Models and Turbulence. Edited by C. Van Atta and M. Rosenblatt. Reprint of the First Edition. VIII, 492 pages. 1975.

Vol. 13: M. Ryan, Hamiltonian Cosmology. VII, 169 pages. 1972.

Vol. 14: Methods of Local and Global Differential Geometry in General Relativity. Edited by D. Farnsworth, J. Fink, J. Porter, and A. Thompson. V, 188 pages.

Vol. 15: M. Fierz, Vorlesungen zur Entwicklungsgeschichte der Mechanik. V, 97 Seiten. 1972.

Vol. 16: H.-O. Georgii, Phasenübergang 1. Art bei Gittergasmodellen. IX, 167 Seiten. 1972.

Vol. 17: Strong Interaction Physics. Edited by W. Rühl and A. Vancura. V, 405 pages. 1973.

Vol. 18: Proceedings of the Third International Conference on Numerical Methods in Fluid Mechanics, Vol. I. Edited by H. Cabannes and R. Temam. VII, 186 pages. 1973.

Vol. 19: Proceedings of the Third International Conference on Numerical Methods in Fluid Mechanics, Vol. II. Edited by H. Cabannes and R. Temam. VII, 275 pages. 1973.

Vol. 20: Statistical Mechanics and Mathematical Problems. Edited by A. Lenard. VIII, 247 pages. 1973.

Vol. 21: Optimization and Stability Problems in Continuum Mechanics. Edited by P. K. C. Wang. V, 94 pages. 1973.

Vol. 22: Proceedings of the Europhysics Study Conference on Intermediate Processes in Nuclear Reactions. Edited by N. Cindro, P. Kulišic and Th. Mayer-Kuckuk. XIV, 329 pages. 1973.

Vol. 23: Nuclear Structure Physics. Proceedings 1973. Edited by U. Smilansky, I. Talmi, and H. A. Weidenmüller. XII, 296 pages. 1973.

Vol. 24: R. F. Snipes, Statistical Mechanical Theory of the Electrolytic Transport of Nonelectrolytes. V, 210 pages. 1973.

Vol. 25: Constructive Quantum Field Theory. The 1973 "Ettore Majorana" International School of Mathematical Physics. Edited by G. Velo and A. Wightman. III, 331 pages. 1973.

Vol. 26: A. Hubert, Theorie der Domänenwände in geordneten Medien. XII, 377 Seiten. 1974.

Vol. 27: R. K. Zeytounian, Notes sur les Ecoulements Rotationnels de Fluides Parfaits. XIII, 407 pages. 1974.

Vol. 28: Lectures in Statistical Physics. Edited by W. C. Schieve and J. S. Turner. V, 342 pages. 1974.

Vol. 29: Foundations of Quantum Mechanics and Ordered Linear Spaces. Advanced Study Institute, Marburg 1973. Edited by A. Hartkämper and H. Neumann. VI, 355 pages. 1974.

Vol. 30: Polarization Nuclear Physics. Proceedings 1973. Edited by D. Fick. IX, 292 pages. 1974.

Vol. 31: Transport Phenomena. Sitges International Schools of Statistical Mechanics, June 1974. Edited by G. Kirczenow and J. Marro. XIV, 517 pages. 1974.

Vol. 32: Particles, Quantum Fields and Statistical Mechanics. Proceedings 1973. Edited by M. Alexanian and A. Zepeda. V, 132 pages. 1975.

Vol. 33: Classical and Quantum Mechanical Aspects of Heavy Ion Collisions. Proceedings 1974. Edited by H. L. Harney, P. Braun-Munzinger, and C. K. Gelbke. VII, 311 pages. 1975.

Vol. 34: One-Dimensional Conductors GPS Summer School Proceedings, 1974. Edited by H. G. Schuster. VII, 371 pages. 1975.

Vol. 35: Proceedings of the Fourth International Conference on Numerical Methods in Fluid Dynamics, 1974. Edited by R. D. Richtmyer. V, 457 pages. 1975.

Vol. 36: R. Gatignol, Théorie Cinétique des Gaz à Répartition Discrète de Vitesses. II, 219 pages. 1975.

Vol. 37: Trends in Elementary Particle Theory. Proceedings 1974. Edited by H. Rollnik and K. Dietz. V, 472 pages. 1975.

Vol. 38: Dynamical Systems, Theory and Applications. Proceedings 1974. Edited by J. Moser. VI, 624 pages. 1975.

Vol. 39: International Symposium on Mathematical Problems in Theoretical Physics. Proceedings 1975. Edited by H. Araki. XII, 562 pages. 1975.

Vol. 40: Effective Interactions and Operators in Nuclei. Proceedings 1975. Edited by B. R. Barrett. XII, 339 pages. 1975.

Vol. 41: Progress in Numerical Fluid Dynamics. Proceedings 1974. Edited by H. J. Wirz. V, 471 pages. 1975.

Vol. 42: H II Regions and Related Topics. Proceedings 1975. Edited by D. Downes and T. L. Wilson. XII, 488 pages. 1975.

Vol. 43: Laser Spectroscopy. Proceedings 1975. Edited by S. Haroche, J. C. Pebay-Peyroula, T. W. Hänsch, and S. E. Harris. X, 466 pages. 1975.

Lecture Notes in Physics

Edited by J. Ehlers, München, K. Hepp, Zürich,
R. Kippenhahn, München, H. A. Weidenmüller, Heidelberg,
and J. Zittartz, Köln
Managing Editor: W. Beiglböck, Heidelberg

63

V. K. Dobrev · G. Mack · V. B. Petkova
S. G. Petrova · I. T. Todorov

Harmonic Analysis

on the n-Dimensional Lorentz Group and Its Application to Conformal Quantum Field Theory

Springer-Verlag
Berlin · Heidelberg · New York 1977

Authors

V. K. Dobrev
V. B. Petkova
S. G. Petrova
I. T. Todorov
Institute of Nuclear Research and Nuclear Energy
Bulgarian Academy of Sciences
Sofia 1113/Bulgaria

G. Mack
II. Institut für Theoretische Physik der Universität Hamburg
Luruper Chausee 149
2000 Hamburg 50/BRD

Library of Congress Cataloging in Publication Data

Main entry under title:

Harmonic analysis on the n-dimensional Lorentz group and its application to conformal quantum field theory.

(Lecture notes in physics ; 63)
Includes bibliographical references.
1. Lorentz transformations. 2. Quantum field theory.
3. Harmonic analysis. I. Dobrev, V. K.
II. Series.
QC174.52.L6H37 530.1'43 77-5339

ISBN 3-540-08150-X Springer-Verlag Berlin Heidelberg New York
ISBN 0-387-08150-X Springer-Verlag New York Heidelberg Berlin

Printing and binding: Beltz Offsetdruck, Hemsbach/Bergstr. 2153/3140-543210

PREFACE

Work on conformal quantum field theory and partial wave expansions led the authors to the necessity to master a number of topics in representation theory and to solve some problems whose coverage in the literature was not quite adequate. It soon became apparent that a systematic study of representations and intertwining operators for the pseudoorthogonal group is desirable, which would place our subject in the context of modern harmonic analysis. That led to Part One of this book. We included in Part Two (following the opportune advice of Klaus Hepp) some recent results on tensor product expansions and related physical applications (see M2 D4 - D6).

In the course of our work on Part One (particularly during the stay of three of us in Princeton) we had the opportunity to consult a number of experts in the field. At different stages we benefited from discussions with F. Gürsey, H. Kraljevic, D. P. Zhelobenko and especially with G. J. Zuckerman. It is a pleasure to express our sincere gratitude to all of them.

One of us (V.K.D.) would like to thank Professor A. S. Wightman for the hospitality extended to him at the Princeton University. Two of us (G.M. and I.T.T.) thank Professor C. Kaysen for his hospitality at the Institute for Advanced Study.

CONTENTS

INTRODUCTION

The generalized Lorentz (or de Sitter) groups O(N,1) are the most important simple noncompact Lie groups used so far in physics. Interest in their representation theory was recently revived, when it was realized that group theoretical methods can be used for a non-perturbative analysis of conformal invariant quantum field theory and its Euclidean Green functions (see, e.g., [M6, M7, M2, T5, D4-D6] and Chapter V of this book.)

The Euclidean conformal group for 2h space-time dimensions is O(2h + 1, 1). (The case 2h = 3 may be of interest in the study of critical phenomena in statistical mechanics, -cf.[P2].)

On the other hand, these groups belong to the family of real (or split) rank one semi-simple Lie groups, which served as a starting point and are the most developed part of the Harish-Chandra theory of induced representations and harmonic analysis on semi-simple Lie groups [H1,H2,W3,W2] . It is not an accident that this theory originated in the work of Bargmann [B1] and Gel'fand and Naimark [G3,G2] who studied the examples of the lowest rank pseudo-orthogonal groups O(2,1) and O(3,1). The representations of the de Sitter group O(4,1) have also been the object of a special investigation (see, e.g., [D3, T1,K3,K7] and references to earlier work cited there). A study of the unitary irreducible representations (UIR) of O(N,1) for arbitrary N was attempted in [H3, O3] using chiefly infinitesimal methods. However, contrary to authors' claims the classification obtained there is not complete, since it omits the so called class I complementary series which incorporates the entire complementary series in the well known special case of the ordinary Lorentz group O(3,1). (It so happens that this omitted series contains precisely the representations used in the physical applications [M2,D6] .) Recently (while the present work was in progress -cf. [D4]), Thieleker [T2,T3] gave a more comprehensive treatment of the representation theory of O(N,1) (and its double covering Spin (N,1)), based on the abstract approach of Harish-Chandra. (We disagree however with his result concerning one of the exceptional series of unitary representations -- see the discussion in section 6. C below).

After this work was completed (and circulated in a preprint form) we became acquainted with a new preprint [G1] on this subject. It contains, in particular, an infinitesimal form of the intertwining operators which is complementary to the global treatment, presented here. An (incomplete) study of the unitary irreducible representations of the general pseudo-orthogonal group O (p,q) is also presented in [F6].

The aim of Part One is to give a global construction of a class of so called type I (or symmetric tensor) representations of O(N,1) and to study associated quantities (characters, intertwining operators), which are suitable for analytic computations that are needed in physical applications. Often the results are extracted from the general theory of Harish-Chandra supplemented by the Knapp-Stein and Schiffmann study of intertwining operators [K3,K4,S1]. (The method of constructing complementary series of unitary representations by exhibiting the range of positivity of intertwining operators was first used for the Lorentz groups of lower dimensions [G2,B3]. The relevance of the work of Knapp and Stein [K3,K4] for conformal quantum field theory was already pointed out by Koller [K5].) We emphasize this point, because it refutes a widespread prejudice among physicists that modern mathematical theories do not provide useful tools for analytical computations. We will feel rewarded, if the present lecture notes will help to bridge the existing gap between mathematicians and physicists studying and applying the theory of group representations.

On the other hand, it should be pointed out that there is no general theory of unitary irreducible representations (IR's) of semi-simple Lie groups; the harmonic analysis on such groups (which is well developed) does not solve the problem, since not all the unitary IR's actually appear in the harmonic expansion of the regular representations. In the case of the generalized Lorentz group, however, there are several simplifying features (in addition to it being of split rank one), which make the classification of all irreducible representations (both unitary and nonunitary) manageable. First of all, the orthogonal groups, SO(n), are multiplicity free; in particular, each elementary representation of $G = SO^{\uparrow}(N,1)$ (to be defined in Chapter I below) contains any IR of the maximal compact subgroup K = SO (N) at most once. To formulate the second property, we need to introduce the notion of the (minimal) parabolic subgroup NMA of G, which (in our case) comprises a group (NM) conjugate to the Euclidean Poincaré group and the (one dimensional) group of dilatations A. (Sec. 1B). The "translation" or alternatively the special conformal transformation subgroup N, which would be in general, just nilpotent, here is actually abelian. As a result, the harmonic analysis on MAN is easily effected, with the help of (Mackey's straight-forward generalization of) Wigner's celebrated study of the Poincaré group. We use this observation to analyze the positivity constraints on the bilinear forms that define the scalar products in the various unitary representation spaces (Secs. 5C, 6C and Appendix C). These bilinear forms (whose kernels play the role of physical propagators in the applications) are obtained (by a possible restriction and a change of normalization) from the Knapp and Stein intertwining operators (Sec. 4). In the physicist language the harmonic analysis on MAN amounts to first taking the ordinary Fourier transform of the intertwining kernel (or of the x-space expression of the conformal propagator for tensor fields) and then expanding the result in $SO(2h-1)_p$ projection operators (that is, projection onto states of definite spin, $SO(2h - 1)_p$ being the stability subgroup of the "momentum" p) (Sec. 5). This analysis also allows us to investigate the Wightman positivity condition for the 2-point function (Sec.5D). Mathematicians follow usually a different path: viz. harmonic analysis on K, which is rather complicated even in the case of the generalized Lorentz group (cf. [T3,T4]).

Besides the Knapp and Stein intertwining operators there are intertwining differential operators which exhibit partial equivalences among some exceptional elementary representations with the same infinitesimal character (Secs. 3A and 6A,B, cf. [Z5,G1]).

It is noted, that one of the four families of (type I) exceptional (integer) points in the representation space is related to the discrete series of analytic representations of the Minkowski space conformal group $SO(2h, 2)/\mathbb{Z}_2$ considered in [G10,O2,R2] . In the special case 2h = 3 the same family is shown to contain the Harish-Chandra discrete series of O(4,1)(Sec.7). The explicit p-space expression for the intertwining kernels at integer points are used to establish some differential identities between these kernels. These identities become important in physical applications, when one goes over to a Laplace transform on G, in order to derive asymptotic expansions. They are responsible for the cancellation of certain kinematical singularities in such expansions, which come from poles of representation functions of second kind. The results centered around the harmonic analysis of the intertwining kernels and their interrelations at integer points, form the core of Part One. For the benefit of the physicist reader, we also review the relevant mathematical background and give explicit realizations of some often used abstract constructions.

The reader who is only interested in the physical applications envisaged so far may omit Sections 2.C, 3.B, 3.C and 7. Section (8.C) contains a synopsis on unitarity, irreducibility and equivalence properties of elementary representations (which can be read independently). Notations for often encountered subgroups of G are collected in Table 1 (Section 1.B).

Chapter IV is devoted to the explicit construction of the tensor product decomposition of two unitary representations of $G_{ex} = O^{\uparrow}(2h+1,1)$ which belong either to the principal series or to a class I (scalar) supplementary series representation. The Clebsch-Gordan kernels for the tensor product expansion under consideration are constructed in Sec. 10A. The Plancherel formula for the Kronecker product of two principal series representations is derived in Sec. 10B for even space-time dimension (integer h). The derivation uses results of part I. The result is extended to supplementary series representations by analytic continuation in Sec. 10D. Some of the results are extended to half integer h in Sec. 10 C; in particular, it is shown that discrete series components do not appear in the tensor product decomposition of two class I representations.

In Sec. 11 we establish some identities among analytically continued Clebsch Gordan kernels at exceptional (integer) points.
The results of Chapter IV extend earlier results derived in [D5] .

An effort was made in the last few years to exploit the conformal invariance of the Gell-Mann - Low limit theory for some Yukawa type interactions in order to obtain non-perturbative information for Green functions and operator products at short distances (see, e.g. [M9, P1,M7,D2,S8,H6,M6,M10,T4,P3,M2,D4,L2,M3,B6,G9,F5, M5,S3,F4,F1,F3,S4,F8,G10,R2,M8,K6,O2]). In particular, Mack [M2] showed for a model of a self-interacting scalar field that the conformal partial wave expansion of Euclidean Green functions allows to diagonalize and solve the set of renormalized dynamical equations [F7,S6] for that model. It was noted [M10,P3,M2] that the so-called bootstrap equations for the 3-point functions imply the existence of real poles in the conformal partial waves as functions of the dimension. The remaining integral equations of the model lead to some factorization properties for the residues at these poles [M2] .

The main purpose of Chapter V is to derive a discrete expansion for Euclidean Green functions and Wightman functions which corresponds to a vacuum operator product expansion in the terminology of ref. [S4] (i.e. an expansion of the vector distribution $\varphi(x_2)\varphi(x_1)|0\rangle$). The derivation is based on the above results on conformal partial wave analysis and on our previous study of the Clebsch-Gordan expansion for the pseudo-orthogonal group. This approach always involves a conjecture about the analyticity (and the asymptotic behavior) of conformal partial waves, which is partly justified by the analysis of the skeleton diagram expansion [M7] . The identities among Clebsch-Gordan kernels at exceptional integer points in the representation space, derived in Chapter IV, are crucial for cancelling fake singularities coming from kinematical factors (Sec. 15.C). As a by-product we verify a positivity condition for the 4-point Wightman function, which was established in a different manner in ref. [M3]. We also present the complete conformal partial wave decomposition of an arbitrary Euclidean n-point vertex function (Sec. 13 E).

We attempt to make the exposition reasonably self-contained and review a number of results of ref. [M2]. This introductory material also contains some new points: one example is the discussion of the $(\varphi^*\varphi)^2$-model in Secs. 12D and 14A. We mention also the explicit expression for a general "basic field" which enters the operator product expansion of a pair of free fields (Sec. 14.C). The derivation of the vacuum expansion is presented in Sec. 15 (and the related Appendix G). We would like to stress the role of the relations between partially equivalent representations of the Euclidean conformal group established in Part One. Sec. 16 contains a discussion - but no ultimate solution - of the rather difficult problem of incorporating crossing symmetry in the present scheme. It closes by a summary of results (Sec. 16.C).

Each part starts with a brief synopsis.

PART ONE

TYPE I REPRESENTATIONS AND INTERTWINING OPERATORS FOR $O^{\uparrow}(2h+1,1)$

SYNOPSIS

We present global realizations of unitary type I (symmetric tensor) representations of the generalized Lorentz group $O(2h+1,1)$, which are suitable for analytic computations needed in Euclidean conformal invariant quantum field theory in 2h dimensions ($2h = 2,3,4,\ldots$) . Whenever possible the results are extracted from Harish-Chandra's general theory of harmonic analysis on noncompact semi-simple Lie groups, supplemented by the use of the Knapp-Stein and Schiffmann intertwining operators. We also introduce intertwining differential operators, which connect elementary representations at partially equivalent integer points. The core of this part consists of a study of the properties and interrelations of all intertwining operators . We exploit the fact that the nilpotent factor N in the Iwasawa decomposition of $G = SO^{\uparrow}(2h+1, 1)$ is commutative (it consists of 2h dimensional special conformal transformations). The commutativity of N makes elementary the harmonic analysis on the so called parabolic subgroup NMA ($M \approx SO(2h)$, A = one dimensional subgroup of dilatations). Among other things, the harmonic expansion of the intertwining kernel allows one to single out the unitary representations of G. It also provides a simple criterion for positivity of the 2-point Wightman functions in a conformal covariant quantum field theory in 2h-dimensional Minkowski space.

I. ELEMENTARY REPRESENTATIONS OF THE PSEUDO-ORTHOGONAL GROUP

1. Group Structure. Preliminaries

1. A The group O(2h + 1, 1) and its Lie algebra

We shall use the notation $N = 2h + 1$. Thus h is integer for odd N and half odd integer for even N (h plays the role of half sum of the restricted positive roots). We consider $h > 1$ in the main body of the paper, since the lowest dimensional cases ($h = 1$ and $h = \frac{1}{2}$) are degenerate (in some sense) and the representation theory of O(3,1) and O(2,1) is well known (see, however, Appendix B where these special cases are briefly discussed).

The group O(N,1) is defined as the set of linear transformations in the real (N + 1)-dimensional vector space which leave invariant the quadratic form

$$\xi^2 = \xi\eta\xi = \xi_1^2 + \ldots + \xi_N^2 - \xi_0^2$$

$$(\eta_{11} = \ldots \eta_{NN} = -\eta_{00} = 1, \ \eta_{AB} = 0 \text{ for } A \neq B \ . \qquad (1.1)$$

It consists of real $(N + 1) \times (N + 1)$ matrices g satisfying

$$^t g\eta g = \eta \qquad (1.2)$$

($^t g$ being the transposed of g). Its identity component $G = SO^{\uparrow}(N,1)$ comprises those matrices g, satisfying (1.2), for which

$$\det g = 1, \quad g^0_{\ 0} \geq 1. \tag{1.3}$$

We shall mostly deal in what follows with the extended group

$$G_{ex} = O^{\uparrow}(2h + 1, 1) = \{ g \epsilon O(2h + 1, 1) \mid g^0_{\ 0} \geq 1 \} \tag{1.4}$$

which includes space reflections.

The Lie algebra $\mathfrak{g}$ of G consists of real (N + 1) x (N + 1) matrices X such that

$${}^t X\eta + \eta X = 0 . \tag{1.5}$$

We can choose a basis $X_{AB} = -X_{BA}$ (A, B = 0, 1, . . . , N) in $\mathfrak{g}$ satisfying the commutation relations

$$[X_{AB}, X_{CD}] = \eta_{AC}X_{BD} + \eta_{BD}X_{AC} - \eta_{AD}X_{BC} - \eta_{BC}X_{AD} \tag{1.6}$$

(A, B, C, D = 0, . . . , N = 2h + 1).

(Note, that in the physical literature one uses more often the "physical generators" J_{AB} related to the "mathematical generators" X_{AB} by $J_{AB} = iX_{AB}$.) A matrix realization of X_{AB} is given by

$$(X_{AB})^C_{\ D} = \eta_{AD}\delta^C_{\ B} - \eta_{BD}\delta^C_{\ A} . \tag{1.7}$$

1. B Subgroups and decompositions

The group G possesses several important subgroups which are listed in the following table:

Subgroup	Characteristic property	Generators
$K = SO(2h+1)$	maximal compact subgroup	X_{ab} $a, b = 1, \ldots, 2h+1$
$A = SO(1,1)$	1-dimensional, non-compact ("dilatations")	$X_{2h+1\,0} = D$
$M = SO(2h)$	centralizer of A in K ($mam^{-1} = a$) ("Euclidean Lorentz group")	$X_{\mu\nu}$ $\mu, \nu = 1, \ldots, 2h$
N	nilpotent, abelian ("special conformal transformations")	$C_\mu = X_{\mu o} - X_{\mu\, 2h+1}$
$\tilde{N}$	nilpotent, abelian ("translations")	$T_\mu = X_{\mu o} + X_{\mu\, 2h+1}$ $(\mu = 1, \ldots, 2h)$
$H = H_{AM} = AH_M$	noncompact ($[h]+1$)-dimensional abelian group of diagonalizable matrices ("Cartan subgroup")	$D, X_{1\,2}, \ldots, X_{2[h]-1\;2[h]}$

TABLE 1: Remarkable Subgroups of the Connected Pseudo-Orthogonal Group $G = SO^{\uparrow}(2h+1, 1)$ ($[h]$ denotes the integer part of h)

According to (1.6) the generators D, C_μ and T_μ appearing in the last column of the Table satisfy the following (non-trivial) commutation relations:

$$[D, C_\mu] = -C_\mu, \quad [D, T_\mu] = T_\mu, \quad \tfrac{1}{2}[T_\mu, C_\nu] = D\delta_{\mu\nu} - X_{\mu\nu}; \qquad (1.6')$$

$$[X_{\lambda\mu}, T_\nu] = \delta_{\lambda\nu} T_\mu - \delta_{\mu\nu} T_\lambda, \quad [X_{\lambda\mu}, C_\nu] = \delta_{\lambda\nu} C_\mu - \delta_{\mu\nu} C_\lambda.$$

The physical Euclidean space momentum operator is given by a hermitian representation of iT_μ.

Let M' be the normalizer of A in K, -i. e., the set of those elements $m' \in K$ for which $m' a m'^{-1} \in A$ for all $a \in A$. It is easily seen that M' is isomorphic to $O(2h)$ and that M is an invariant subgroup of M'. The finite group

$$W = W(G, A) = M'/M \tag{1.8}$$

is called the <u>Weyl group</u> for the pair (G, A) (or the <u>restricted Weyl group</u>). It has two elements $W = \{1, w\}$. The nilpotent subgroups N and $\tilde{N}$ in Table 1 are conjugate under the Weyl transformation

$$wNw^{-1} = \tilde{N}. \tag{1.9}$$

Whenever needed, we shall choose the following representative of w in M':

$$w = \exp(\pi X_{2h\ 2h+1}) \tag{1.10}$$

i. e. rotation by π in the $(2h, 2h+1)$-plane. In what follows we shall denote the elements of K, N, A, M, $\tilde{N}$ by k, n, a, m, $\tilde{n}$, respectively.

The Iwasawa decomposition (see, e. g., [W3, W2]). Every element g of G may be presented (uniquely) in the factorized form

$$g = kna, \tag{1.11}$$

or (equivalently)

$$g = \tilde{n}ak \tag{1.12}$$

(the factors k and a in (1.11) and (1.12) being, in general, different). The dimension of the abelian factor A in the Iwasawa decomposition $G = KNA$ of a semi-simple Lie group G is called <u>split rank</u> of G.

The order of the factors in both versions of the Iwasawa decomposition is a matter of convenience. Our choice of the first factor in both cases is related to the imbedding of the elementary induced representations of G in the left regular representation of this group (see Section 2). In this and the following chapter we shall use the form (1.11). The form (1.12) will be used in Chapter III in the study of the discrete series of unitary representations of $O^{\uparrow}(4,1)$.

The (Gel'fand-Naimark) Bruhat decomposition (see, e.g., [G3, K3, K4, W3, W2]). Almost every element of G (more precisely, every $g \in G$ which does not belong to the lower dimensional manifold wNAM) can be written in a unique way as a product

$$g = \tilde{n}nam. \tag{1.13}$$

Finally we comment on the last line of Table 1. There is an important difference between the groups $SO^{\uparrow}(N,1)$ for odd and even N. For odd N (integer h) there is only one Cartan subgroup, up to conjugacy within the group. It is non-compact and is reproduced in Table 1. For even N (half integer h) there is, in addition, a compact Cartan subgroup, generated by $X_{12}, X_{34}, \ldots, X_{N-1N}$. It follows that the set of elliptic elements (matrices with eigenvalues of modulus 1) has the same dimension as G for even N, but not for odd N. This difference has a bearing on the structure of the representations of G: only for even N there exists a discrete series of unitary representations of G (see Sections 7 and 8.B, below).

1.C The compactified Euclidean space as a homogeneous space of G.

The "parabolic subgroup" NAM (which appears in the Bruhat decomposition) plays a privileged role in the Harish-Chandra construction of induced representations of G. It is notable that the (group theoretically) distinguished homogeneous space

$$G/NAM \approx K/M \approx S^{2h} \tag{1.14}$$

(isomorphic to the unit sphere S^{2h} in 2h + 1 dimensions) is also relevant to physics (see e.g.[A1]). It can be identified with the conformal compactification of the 2h-dimensional Euclidean space $\mathbb{X} = \mathbb{R}^{2h}$ -- the carrier space of Euclidean (quantum) fields. (S^{2h} is obtained from $\mathbb{X}$ by adding a single point, say ∞; in the group theoretical language that is the equivalence class of the non-trivial element (1.10) of the restriced Weyl group.)

The group G acts in a natural way (by left translation) on the homogeneous space (1.14). In particular, there is a well defined local action of G on the vector space $\mathbb{X}$ which can be identified with the manifold of right cosets $\tilde{n}NMA$ ($\tilde{n} \in \tilde{N}$). Because of uniqueness of the Bruhat decomposition (1.13) for $g \in G \setminus wNAM$ (see , e.g. , Theorem 1.2.1.2. of [W3]) there is a one to one correspondence between the cosets $x \in \mathbb{X}$ and the elements $\tilde{n} \in \tilde{N}$ such that $x = \tilde{n}NMA$. We shall use the notation $\tilde{n}_x$ for

the (uniquely determined) element of $\tilde{N}$ corresponding to x $(\in \mathbb{X})$
We endow $\mathbb{X}$ with the structure of a real vector space in such a way that

$$\tilde{n}_{x_1} \tilde{n}_{x_2} = \tilde{n}_{x_1 + x_2} \tag{1.15a}$$

That gives us an additive parametrization of the abelian subgroup $\tilde{N}$ of G. The parabolic subgroup $\tilde{N}MA$ of G plays the role of automorphism group of $\mathbb{X}$. In particular we have

$$\tilde{n}_x y = x + y \tag{1.15b}$$

The stability subgroup of the zero vector $x = 0$ (corresponding to the unit element of $\tilde{N}$) is the parabolic subgroup NAM of G.

The homogeneous space (1.14) is isomorphic to the set of isotropic rays in $\mathbb{R}^{2h+2}$ (with metric given by (1.7)):

$$G/NAM \approx \mathbb{K}_+ / \mathbb{R}_+ \tag{1.14'}$$

where $\mathbb{R}_+$ is the multiplicative group of positive reals and

$$\mathbb{K}_+ \equiv \mathbb{K}_+^{2h+1,1} \; (\approx G/NM) = \{\xi \in \mathbb{R}^{2h+2} ; \xi_0 > 0, \; \xi\eta\xi = \\ = \xi_1^2 + \dots + \xi_{2h+1}^2 - \xi_0^2 = 0\}$$

[See remark in Section 2. B below, where the components of ξ are identified as linear combinations of the matrix elements of g $(\in G)$.] The components of $x \in \mathbb{X}$ are expressed in terms of the ξ's by

$$x_\mu = (\xi_0 + \xi_{2h+1})^{-1} \xi_\mu \; , \; \text{for } \xi_0 + \xi_{2h+1} \neq 0 \tag{1.16}$$

The left translation by g on the homogeneous space (1.14') generate the natural action $g: \zeta \to g\zeta$ of G on $\mathbb{K}_+$, which gives rise to a well defined transformation law on the x's . We deduce that the automorphism group of $\mathbb{X}$ is the parabolic subgroup $\tilde{N}AM$ (conjugate to NAM). The action of the translations from $\tilde{N}$ is already exhibited in (the second equation) (1.15). The dilatations $a \in A$ and the rotations $m \in M$ act as homogeneous transformations on $\mathbb{X}$:

$$ax = |a| x, \qquad |a| > 0 \tag{1.17a}$$

$$(mx)_\mu = m_{\mu\nu} x_\nu \qquad (\mu = 1, \ldots 2h) \tag{1.17b}$$

(summation is to be carried out over repeated Greek indices from 1 to 2h). For special conformal transformations $n \in N$ we can only speak about their infinitesimal action on $\mathbb{X}$, since for every $x \in X$ there exists an $n \in N$ which would carry it to ∞. For $n = n_\varepsilon$ in the neighborhood of the identity of N ($\varepsilon = \varepsilon_1, \ldots, \varepsilon_{2h}$ being an infinitesimal 2h-vector), we have the infinitesimal law

$$n_\varepsilon : x \to x'(\varepsilon), \quad x'_\mu = x_\mu + (x^2 \delta_{\mu\nu} - 2 x_\mu x_\nu) \varepsilon_\nu + O(\varepsilon^2) \quad . \tag{1.18}$$

We shall derive a global form of this law in the next subsection starting from an explicit realization of the Bruhat decomposition.

1.D Matrix realization of various subgroups of G. Construction of the Bruhat decomposition

In what follows it will be useful to have an explicit realization of the different subgroups in Table 1 which appear in the decompositions (1.11-13), in terms of (2h + 2) x (2h + 2) matrices.

The matrices $\tilde{n}$ and n will be parametrized by the corresponding 2h-dimensional vectors $\mathbf{x}(\in \mathbf{X})$ and b, respectively. Using the matrix realization (1. 7) of the generators we obtain the following expressions for the matrices $\mathbf{x}T \equiv \sum_{\nu=1}^{2h} \mathbf{x}_\nu T_\nu$ and bC:

$$(\mathbf{x}T)^\lambda_{\ \mu} = 0 \qquad (\mathbf{x}T)^{2h+1}_{\ \ \mu} = (\mathbf{x}T)^0_{\ \mu} = \mathbf{x}_\mu$$

$$-(\mathbf{x}T)^\lambda_{\ 2h+1} = (\mathbf{x}T)^\lambda_{\ 0} = \mathbf{x}^\lambda \ (=\mathbf{x}_\lambda), \ \lambda, \mu, = 1, \ldots, 2h,$$

$$(\mathbf{x}T)^A_{\ B} = 0 \text{ for } A, B = 2h+1, 0$$

or, in a more compact notation (in which the 0-th row and column appear in 2h + 2-nd place),

$$\mathbf{x}T = \left(\begin{array}{c|cc} O_{2h} & -{}^t\mathbf{x} & {}^t\mathbf{x} \\ \hline \mathbf{x} & & \\ \mathbf{x} & \multicolumn{2}{c}{O_2} \end{array}\right) \tag{1. 19a}$$

where O_k stands for a k x k matrix with zero elements, $\mathbf{x} = (\mathbf{x}_1, \ldots, \mathbf{x}_{2h})$ is a row vector, and ${}^t\mathbf{x}$ is the corresponding column vector (cf. [K5]); similarly,

$$bC = \left(\begin{array}{c|cc} O_{2h} & {}^tb & {}^tb \\ \hline -b & & \\ b & \multicolumn{2}{c}{O_2} \end{array}\right) . \tag{1. 19b}$$

Using further the nilpotent character of the matrices (1.19) (which obey $(xT)^3 = (bC)^3 = 0$), we obtain

$$\tilde{n}_{\mathbf{x}} = e^{\mathbf{x}T} = \begin{pmatrix} \mathbb{1}_{2h} & -{}^t\mathbf{x} & {}^t\mathbf{x} \\ \mathbf{x} & 1-\frac{1}{2}\mathbf{x}^2 & \frac{1}{2}\mathbf{x}^2 \\ \mathbf{x} & -\frac{1}{2}\mathbf{x}^2 & 1+\frac{1}{2}\mathbf{x}^2 \end{pmatrix} \tag{1.20a}$$

($\mathbb{1}_{2h}$ being the 2h-dimensional unit matrix),

$$n_b = e^{bC} = \begin{pmatrix} \mathbb{1}_{2h} & {}^tb & {}^tb \\ -b & 1-\frac{1}{2}b^2 & -\frac{1}{2}b^2 \\ b & \frac{1}{2}b^2 & 1+\frac{1}{2}b^2 \end{pmatrix} . \tag{1.20b}$$

Analogously, we find that the matrix expression for a dilatation a is

$$a = e^{\alpha D} = \begin{pmatrix} \mathbb{1}_{2h} & 0 & 0 \\ 0 & \mathrm{ch}\alpha & \mathrm{sh}\alpha \\ 0 & \mathrm{sh}\alpha & \mathrm{ch}\alpha \end{pmatrix} = \begin{pmatrix} \mathbb{1}_{2h} & 0 & 0 \\ 0 & \frac{|a|^2+1}{2|a|} & \frac{|a|^2-1}{2|a|} \\ 0 & \frac{|a|^2-1}{2|a|} & \frac{|a|^2+1}{2|a|} \end{pmatrix} \tag{1.21}$$

$|a| = e^{\alpha}\ (> 0)$.

In the same notation

$$m = \begin{pmatrix} m_{\mu\nu} & 0 & 0 \\ 0 & 1 & 0 \\ 0 & 0 & 1 \end{pmatrix}, \quad (m_{\mu\nu}) \epsilon\, SO(2h), \tag{1.22}$$

$$k = \begin{pmatrix} k_{\mu\nu} & k_{\mu\, 2h+1} & 0 \\ k_{2h+1\,\nu} & k_{2h+1\, 2h+1} & 0 \\ 0 & 0 & 1 \end{pmatrix}, (k_{ab}) \epsilon\, SO(2h+1). \tag{1.23}$$

Multiplying the matrices (1. 20) - (1. 22) we obtain the following expression for a group element g of the form (1. 13)

$$g = \tilde{n}_x n_b a m = \begin{pmatrix} \mathbf{1}_{2h} & -x & x \\ x & 1-\frac{1}{2}x^2 & \frac{1}{2}x^2 \\ x & -\frac{1}{2}x^2 & 1+\frac{1}{2}x^2 \end{pmatrix} = \begin{pmatrix} m^\mu{}_\nu & |a| b^\mu & |a| b^\mu \\ -m^\sigma_\nu b_\sigma & \frac{|a|^2(1-b^2)+1}{2|a|} & \frac{|a|^2(1-b^2)-1}{2|a|} \\ m^\sigma_\nu b_\sigma & \frac{|a|^2(1+b^2)-1}{2|a|} & \frac{|a|^2(1+b^2)+1}{2|a|} \end{pmatrix} =$$

$$= \begin{pmatrix} m^\mu{}_\nu + 2x^\mu b_\sigma m^\sigma{}_\nu & |a| b^\mu + (|a| b^2 - \frac{1}{|a|}) x^\mu & |a| b^\mu + (|a| b^2 + \frac{1}{|a|}) x^\mu \\ x_\sigma m^\sigma{}_\nu + b_\sigma m^\sigma{}_\nu (x^2-1) & |a|(xb) + \frac{|a|^2 + (|a|^2 b^2 - 1)(x^2-1)}{2|a|} & |a|(xb) + \frac{|a|^2 + (|a|^2 b^2 + 1)(x^2-1)}{2|a|} \\ x_\sigma m^\sigma{}_\nu + b_\sigma m^\sigma{}_\nu (x^2+1) & |a|(xb) + \frac{(|a|^2 + (|a|^2 b^2 - 1)(x^2+1)}{2|a|} & |a|(xb) + \frac{|a|^2 + (|a|^2 b^2 + 1)(x^2+1)}{2|a|} \end{pmatrix}$$

(1. 24)

The condition that a matrix $g \epsilon G$ can be written in the form (1. 24) is

$$d(g) \equiv \tfrac{1}{2} (g^{2h+1}{}_{2h+1} - g^{2h+1}{}_{0} - g^{0}{}_{2h+1} + g^{0}{}_{0}) > 0 \ . \qquad (1.\ 25)$$

(The use of upper and lower indices is only necessary when the index 0 is involved; for $\mu, \nu = 1, \ldots 2h$ we have $m^{\mu}{}_{\nu} = m_{\mu\nu}$, $b^{\mu} = b_{\mu}$ etc.) We note that d(g) is always nonnegative (for $g \epsilon G$) and vanishes for g of the form wnam.

Using (1. 24) we find

$$|a| \equiv |a|_{B}(g) = \frac{1}{d(g)} \ , \qquad (1.\ 26a)$$

$$x^{\mu} \equiv x^{\mu}{}_{B}(g) = \frac{1}{2d(g)} \left(g^{\mu}{}_{0} - g^{\mu}{}_{2h+1} \right), \qquad (1.\ 26b)$$

$$2b^{\mu} \left(\equiv 2b^{\mu}{}_{B}(g) \right) = (g^{0}{}_{0} - g^{2h+1}{}_{0}) g^{\mu}{}_{2h+1} + (g^{2h+1}{}_{2h+1} - g^{0}{}_{2h+1}) g^{\mu}{}_{0} \ , \qquad (1.\ 26c)$$

$$m^{\mu}{}_{\nu} \left(\equiv m^{\mu}_{B\nu}(g) \right) = g^{\mu}{}_{\nu} - \frac{(g^{\mu}{}_{0} - g^{\mu}{}_{2h+1})(g^{0}{}_{\nu} - g^{2h+1}{}_{\nu})}{2d(g)} \ . \qquad (1.\ 26d)$$

(The subscript B is to remind us that the quantities $|a|_{B}$, $x^{\mu}{}_{B}$, etc. are determined from the Bruhat decomposition.)

We define the transformation $x \xrightarrow{g} x'$ from

$$g\tilde{n}_{x} = \tilde{n}_{x'} \, n(g,x) a(g,x) m(g,x). \qquad (1.\ 27)$$

This leads to the transformation law (1. 15) (1. 17) under the automorphism group $\widetilde{N}AM$ of $\mathbf{X}$ and to the law

$$\mathbf{x}' = \frac{\mathbf{x} + \mathbf{x}^2 b}{1 + 2b\mathbf{x} + b^2 \mathbf{x}^2} \qquad (1.\ 28)$$

under special conformal transformations (which agrees with the infinitesimal rule (1. 18) for $b = \varepsilon \to 0$). We notice that the denominator in (1. 28) coincides with the function $d(n_b \tilde{n}_{\mathbf{x}})$ defined by (1. 25). It is useful to observe that the special conformal transformation can be expressed in terms of a translation $\tilde{n}$ and the reflection R of the first 2h + 1 axes (R is equal to the Weyl transformation w defined in (1. 10) followed by a reflection $\xi_i \to -\xi_i$, i = 1, . . . , 2h-1; for odd 2h R is a proper G transformation and could be taken as a representative of the non-trivial element of W instead of (1. 10)). Using the decomposition (1. 27) of $R\tilde{n}_{\mathbf{x}}$ we deduce the following transformation law under <u>conformal inversion</u>:

$$R\mathbf{x} = -\frac{\mathbf{x}}{\mathbf{x}^2}\ ; \qquad wx = \frac{\theta x}{x^2} \qquad (1.\ 29)$$

It is easy to verify that $n_b = R\tilde{n}_{-b}R^{-1}$, θ is reflection of the 2h-th axis.

Let us return to Eq. (1.27). It will be convenient for applications in part II of this book to introduce some special notation and state a few identities. We define $p(x,g) \in MAN$ by

$$g^{-1}\tilde{n}_x = \tilde{n}_{x'}p(x,g)^{-1} \qquad x' = g^{-1}. \tag{1.27a}$$

Thus $p(x,g^{-1})^{-1} = n(g,x)a(g,x)m(g,x)$ in the notation of (1.27). From (1.27a) we deduce the cocycle condition

$$p(x,g_1g_2) = p(x,g_1)\, p(g_1^{-1}x, g_2) \tag{1.27b}$$

The following special cases are immediate from the definition

$$p(x,\tilde{n}) = 1 \quad \text{for } \tilde{n} \in \tilde{N}\ ; \quad p(x,ma) = ma \quad \text{for } ma \in MA \tag{1.27c}$$

For

$$p_x \equiv p\ (x,w) \tag{1.27d}$$

Eq. (1.24) yields the explicit formula

$$p_x = m_x a_x n_{\theta x} \quad \text{with } m_x = \left(-\mathbb{1} + 2\frac{xx^t}{x^2}\right)I_s\ , \quad |a(x)| = x^2 \tag{1.27e}$$

We used matrix notation; x resp. x^t denote the column resp. row vector with entries $x^1 \ldots x^{2h}$. The following identities are immediate consequences

$$m_{x+y}\, m_{wy} = m_x\, m_{wx+wy}\ ; \quad a_{x+y}\, a_{wy} = a_x\, a_{wx+wy} \tag{1.27f}$$

and

$$\begin{aligned} &p(x,n_{-Ry}) = h_x\, h_{wx-wy}\, n_{Ru} = n_{R(y-x)}\, h_x\, h_{wx-wy} \\ &\text{where } h_x = m_x a_x \quad \text{etc. and } u = y + w(wx-wy) \end{aligned} \tag{1.27g}$$

1. E Relationship between the Bruhat and the Iwasawa decomposition. The Haar measure

The functions $k(g)$, $b_I(g)$ and $|a|_I(g)$ corresponding to the Iwasawa decomposition (1. 11) can be found in a similar way as the expressions (1. 26). In order to exhibit the relation between the representations (1. 11) and (1. 13) it sufficies to write down the Iwasawa decomposition of $\tilde{n}_x$ and the Bruhat decomposition of k.

We have

$$k(\tilde{n}_x) = \begin{pmatrix} \delta^\mu{}_\nu - 2\dfrac{x^\mu x_\nu}{1+x^2} & -\dfrac{2x^\mu}{1+x^2} & 0 \\ \dfrac{2x_\nu}{1+x^2} & \dfrac{1-x^2}{1+x^2} & 0 \\ 0 & 0 & 1 \end{pmatrix} \equiv k_x \qquad (1.30a)$$

$$b_I(\tilde{n}_x) = x \qquad (1.30b)$$

$$|a|_I(\tilde{n}_x) = \frac{1}{1+x^2} \equiv |a(x)| \, . \qquad (1.30c)$$

To prove (1. 30) it is sufficient to verify that

$$k_x n_x a(x) = \tilde{n}_x , \qquad (1.31)$$

which is a straightforward exercise in matrix multiplication.

In order to write down the Bruhat decomposition of k, we first note that each $k \in K$ can be written in the form

$$k = k_x m \quad \text{with } x^2 = \frac{1-k_{2h+1\ 2h+1}}{1+k_{2h+1\ 2h+1}} \qquad (1.32)$$

where k_x is given by (1. 30a) and $m \in M$. Indeed, k_x is nothing but a rotation in the $(x, 2h+1)$ plane by an angle

$$\theta = \arccos \frac{1-x^2}{1+x^2} \tag{1. 33}$$

and (1. 32) is a standard decomposition for SO(2h+1). Since the Bruhat decomposition of m is trivial ($m_B = m$) the decomposition of an arbitrary $k \in K$ is reduced to the decomposition of k_x, which is given by (1. 13) with

$$x(k_x) = x, \quad m_B(k_x) = 1, \tag{1. 34}$$

$$|a|_B(k_x) = 1 + x^2, \quad b_B(k_x) = \frac{-x}{1+x^2}. \tag{1. 35}$$

Finally, we shall write down the invariant measure on G in terms of the invariant measures of its subgroups in the Iwasawa and the Bruhat decomposition. Since the group G is unimodular, its Haar measure is both left and right invariant; the measures of the non-unimodular factors will be chosen to be left invariant.

Let dk be the Haar measure on K normalized by

$$\int_K dk = 1. \tag{1. 36}$$

Set further

$$da = \frac{d|a|}{|a|} \; (=d\alpha), \quad dn_b = db_1 \ldots db_{2h} \equiv db. \tag{1. 37}$$

The left invariant measure on NA can be written as

$$d(na) = |a|^{2h} dn_b da. \tag{1.38}$$

In order to check (left) dilatation invariance of (1.38), we note that

$$an_b = n_{b/|a|}a. \tag{1.39}$$

Then the Haar measure on G has the form

$$dg = d\kappa \, db \, da \tag{I.40}$$

(for g given by (I.II)). To see that we note that dg has to be both left and right invariant (since the group G is unimodular) and the expression (I.40) is fixed (up to overall normalization) by the requirement that it is left invariant under K and right invariant under NA.

In order to express dg in terms of the factors of the Bruhat decomposition (I.13), we first note that the right invariant measure on the parabolic subgroup NAM is

$$d_R(n_b am) = db \, da \, dm$$

where dm is the normalized Haar measure on M. That follows from the semi-direct product structure of NAM = N$\circledS$(A$\otimes$M) (N being an invariant subgroup). Using again that dg is both left and right invariant, we obtain

$$dg = dx \, db \, da \, dm \tag{I.41}$$

where $dx = dx_1 \ldots dx_{2h}$ ($= d\tilde{n}_x$).

It is also worth noting that the invariant measure on $K/_M$ obtained from the decomposition (I.32) is

$$d\kappa_x = \frac{\Gamma(2h)\; dx}{\pi^h \Gamma(h)(1+x^2)^{2h}}. \tag{I.42}$$

2. Induced representations. Definition and various realizations

2. A Synopsis on the irreducible representations of the orthogonal group [29, 30]

Here we recall some basic facts (and fix notation conventions) on the irreducible representations of the compact orthogonal group O(n) to which we shall refer in the rest of this section. Some additional information on these representations is assembled for later use in Appendix A.

The finite dimensional irreducible representations (IR) of a compact Lie group are completely characterized by their highest weight (see [30]). One has to distinguish the type of highest weights of SO(n) for the cases of even and odd dimensions.

The highest weight for the orthogonal group of rank ν (that is, for SO(2ν) or SO(2ν + 1)) is $\ell = (\ell_1, \ldots, \ell_\nu)$ where

$$|\ell_1| \leq \ell_2 \leq \ldots \leq \ell_\nu \qquad \text{for } SO(2\nu) \qquad (2.1a)$$

$$(\nu \geq 1)$$

$$0 \leq \ell_1 \leq \ell_2 \leq \ldots \leq \ell_\nu \qquad \text{for } SO(2\nu+1) . \qquad (2.1b)$$

The numbers ℓ_i are all integers for single valued IR of SO(n) and all half integers for the double valued IR (i. e. for the faithful IR of the double covering Spin (n) of SO(n)). The dimension of the IR $(\ell_1, \ldots, \ell_{[h]})$ of SO(2h) is given by

$$d_{2h}(\ell) \equiv d(\ell_1,\ldots,\ell_h) = \prod_{1\le i\le j\le h} \frac{(n_i+n_j)(n_j-n_i)}{(2h-i-j)(j-i)} \quad \text{for } h=2,3,\ldots; \tag{2.2a}$$

$$d_2(\ell) = 1 \quad \text{for } h = 1;$$

$$d_{2h}(\ell) \equiv d(\ell_1,\ldots,\ell_{h-\frac{1}{2}}) = \prod_{i=1}^{h-\frac{1}{2}} \frac{2n_i}{2(h-i)} \prod_{i<j} \frac{(n_i+n_j)(n_j-n_i)}{(2h-i-j)(j-i)} \tag{2.2b}$$

$$\text{for } h = \frac{3}{2}, \frac{5}{2}, \ldots,$$

where

$$n_{[h]+1-j} = \ell_{[h]+1-j} + h - i \qquad i = 1,\ldots,[h]. \tag{2.3}$$

(The vector $(h-[h],\ldots,h-1)$ added to the highest weight in (2.3) is half the sum of positive roots of SO(2h).) The (second order) Casimir operator is

$$\Omega(\ell_1\ldots\ell_{[h]}) = \tfrac{1}{2}X_{\mu\nu}X^{\nu\mu} = \sum_{i=1}^{[h]} [n^2_{[h]+1-i} - (h-i)^2] = \sum_{i=1}^{[h]} \ell_{[h]+1-i}(\ell_{[h]+1-i}+2h-2i). \tag{2.4}$$

(We are using here the symbol $X_{\mu\nu}$ for the representation ℓ of the generators X, introduced in Section 1.A.)

The highest weight characterization of the IRs of SO(n) is particularly useful in studying their SO(n-1) content [G5]. A (unitary) IR $\ell' = (\ell'_1,\ldots,\ell'_{\nu-1})$ of $SO(2\nu-1)$ is contained in the IR

$\ell = (\ell_1, \ldots, \ell_\nu)$ of $SO(2\nu)$ iff

$$|\ell_1| \leq \ell'_1 \leq \ell_2 \leq \cdots \leq \ell'_{\nu-1} \leq \ell_\nu \tag{2.5}$$

The IR $\ell' = (\ell'_1, \ldots, \ell'_\nu)$ of $SO(2\nu)$ is contained in the IR $\ell = (\ell_1, \ldots, \ell_\nu)$ of $SO(2\nu+1)$ iff

$$-\ell_1 \leq \ell'_1 \leq \ell_1 \leq \ell'_2 \leq \cdots \leq \ell'_\nu \leq \ell_\nu. \tag{2.6}$$

Each IR of SO(n-1) is contained in a given IR of SO(n) at most once.

Let now O(n) be the full orthogonal group, including space reflections π such that

$$\pi^2 = {}^t\pi\pi = 1, \qquad \det \pi = -1. \tag{2.7}$$

Let $D^\ell(\Lambda)$ be an IR of SO(n) acting in the (complex) vector space $\mathcal{V}^\ell$. For a given reflection $\pi \in O(n)$ (satisfying (2.7) we define the mirror image of D^ℓ as the representation

$$D^{\pi\ell}(\Lambda) \equiv D^\ell(\pi\Lambda\pi^{-1}) \tag{2.8}$$

acting in the same vector space $\mathcal{V}^\ell$ (we note that $\pi\Lambda\pi^{-1} \in SO(n)$ for any $\Lambda \in SO(n)$, so that the right hand side of (2.8) is well defined.).

The following facts about mirror images are well known (see e.g. [W4] or Section 114 of ref. [Z2]).

(i) The mirror image of an IR $\ell = (\ell_1, \ldots, \ell_\nu)$ of $SO(2\nu)$ is equivalent to $\pi\ell = (-\ell_1, \ell_2, \ldots, \ell_\nu)$.

(ii) The mirror image of any IR ℓ of SO($2\nu + 1$) is equivalent to ℓ.

Thus any IR of SO($2\nu + 1$) can be extended to an IR of O($2\nu + 1$), while an IR $\ell = (\ell_1 \ldots \ell_\nu)$ of SO(2ν) can be extended to an IR of O(2ν) if and only if $\ell_1 = 0$.

Having in mind the applications envisaged in [M2,D4], we shall be particularly interested in the symmetric traceless tensor IR (which will be also called type one representations) with highest weight

$$\ell_I = (0, \ldots, 0, \ell). \tag{2.9}$$

The representation of this type can be realized in the vector space $\mathcal{V}^\ell$ of all homogeneous polynomials of degree ℓ

$$f(z) = f_{\mu_1 \ldots \mu_\ell} z_{\mu_1} \ldots z_{\mu_\ell} \tag{2.10}$$

on the complex light cone

$$\mathbb{K}_n = \{z \in \mathbb{C}^n;\ z^2 = z_1^2 + \ldots + z_n^2 = 0\} \tag{2.11}$$

(cf. [T6,O1,Z1,B2]). Each such function allows a unique harmonic (homogeneous polynomial) extension to the whole n-dimensional complex space $\mathbb{C}^n$ (see [B2] and Appendix A). The group action on vectors of the type (2.10) is given by

$$[D^\ell(\Lambda)f](z) = f(\Lambda^{-1} z). \tag{2.12}$$

It follows from (2.5) (2.6) and (2.9) that the type I representations of SO(n) are decomposed into only type I representations of SO(n-1). Concerning the expression for the convolution of two tensors in the homogeneous polynomial formalism see Appendix A.5.

2. B Covariant vector valued functions on G. Definition of the induced representations

In this and the following subsections we will give three alternative realizations of the induced representations of $G_{(ex)}$ (induced by the parabolic subgroup MNA). We start with the most abstract and general one.

Let $\mathcal{V}^{\ell}$ be the (finite dimensional) Hilbert space in which the unitary IR D^{ℓ} of M = SO(2h) is realized. Let further c be an arbitrary complex number and write

$$\chi = [\ell, c] = (\ell, -h-c) \qquad (2.13)$$

(as it will become clear in the sequel the set of numbers in the parentheses corresponds to the generalized highest weight of an elementary representation of $SO^{\uparrow}(2h+1, 1)$). Consider the space $\mathcal{C}_{\chi}$ of infinitely differentiable functions f on G with values in $\mathcal{V}^{\ell}$, satisfying the covariance condition

$$f(gnam) = |a|^{h+c} D^{\ell}(m)^{-1} f(g). \qquad (2.14)$$

The representation $\mathcal{T}(g) = \mathcal{T}^{\chi}(g)$ of G, induced by the finite dimensional representation $|a|^{-h-c} D^{\ell}(m)$ of MNA, is defined by

$$[\mathcal{T}(g) f](g') = f(g^{-1}g'), \quad g, g' \in G, \quad f \in \mathcal{C}_{\chi} . \qquad (2.15)$$

(It differs from the left regular representation of G by the domain $\mathcal{C}_\chi$ of the operators $\mathcal{T}(g)$ which is characterized by the covariance condition (2.14).) The representations so constructed are called elementary (induced) representations of G.

Remark: Let us exhibit the relation between this standard mathematical construction and the so called manifestly covariant formalism of ref. [M7, T5]. In order to simplify the discussion we shall restrict ourselves to the case of a scalar representation ($\ell = 0$).

It follows from the covariance property (2.14) that a function $f \in \mathcal{C}_\chi$ is completely fixed by its values on the homogeneous space (1.14). The points of this space can be identified with isotropic ("light-like") rays in $\mathbb{R}^{2h+2}$ parametrized by the (homogeneous) coordinates

$$\xi^A = \tfrac{1}{2}\left(g^A{}_0 - g^A{}_{2h+1}\right), \qquad A = 1, \ldots, 2h+1, 0. \tag{2.16}$$

It is a simple consequence of the orthogonality condition (1.2) that

$$\xi^2 = \xi_A \xi^A = \xi^A \eta_{AB} \xi^B = 0. \tag{2.17}$$

Furthermore, using (1.21) (1.20b), (1.22), and (2.16), we obtain

$$\xi(ga) = \frac{1}{|a|}\,\xi(g), \quad \xi(gn) = \xi(gm) = \xi(g). \tag{2.18}$$

Thus the isotropic vectors ξ (2.16) are in one-to-one correspondence with the equivalence classes $gMN \in G/MN$ and can be identified with them. On the other hand, a function F(g), satisfying (2.14) with $\ell = 0$, only depends on these equivalence classes, so that

$$\mathfrak{f}(g) = \mathfrak{f}(gMN) \equiv \mathfrak{f}(\xi). \tag{2.19}$$

Moreover, because of the first equation (2.18) and (2.14), $\mathfrak{f}(\xi)$ is a homogeneous function of degree -h-c of ξ. The representation (2.15) assumes the form

$$\left[\mathcal{T}^{[0,c]}(g)\mathfrak{f}\right](\xi) = \mathfrak{f}(g^{-1}\xi). \tag{2.20}$$

The vector (2.16) satisfies all properties of the isotropic vector ξ which appears in (1.16) and can therefore be used as an intermediate variable in the passage to the noncompact picture.

2.C The compact picture. K-content of the elementary representations

By the Iwasawa decomposition every element g of G can be written in the form (1.11). The covariance property (2.14) then implies that each (vector valued) function $\mathfrak{f}(g) \in \mathcal{C}_\chi$ can be written in the form

$$\mathfrak{f}(kna) = |a|^{h+c}\mathfrak{f}(k) \tag{2.21}$$

and is thus completely determined by its values on K. Conversely, every smooth function $\mathfrak{f}(k)$ on K satisfying the covariance condition

$$\mathfrak{f}(km) = D^{\ell}(m^{-1})\mathfrak{f}(k) \tag{2.22}$$

can be extended via (2. 21) to an element of $\mathcal{C}_\chi$ (considered as a function on G). Hence, we may identify each $\mathcal{C}_\chi$ with a standard space $C(K, \mathcal{V}^\ell)$ of covariant functions on K. The space $C(K, \mathcal{V}^\ell)$ is independent of c. In fact due to (2. 22) $f(k)$ is completely determined by its values on the unit sphere (1. 14). Explicitly, according to (1. 32) each $k \in K$ is decomposed uniquely in the form

$$k = k_{\hat{\xi}} M \text{ with } k_{\hat{\xi}} = \left(\begin{array}{c|cc} \delta^\mu{}_\nu - \dfrac{\hat{\xi}^\mu \hat{\xi}_\nu}{1+\hat{\xi}_{2h+1}} & -\hat{\xi}^\mu & 0 \\ \hline \hat{\xi}_\nu & \hat{\xi}_{2h+1} & 0 \\ 0 & 0 & 1 \end{array}\right) \qquad (2.\ 23)$$

$$\hat{\xi}^2 = \sum_{a=1}^{2h+1} \hat{\xi}^a \hat{\xi}_a = 1;$$

because of (2. 22) $f(k)$ is fixed (for a given D^ℓ) by its value for $k = k_{\hat{\xi}}$.

Let k_g and $a = a(k, g)$ be determined from the Iwasawa decomposition of $g^{-1}k$:

$$g^{-1}k = k_g na.$$

Then we define the compact picture realization $\tilde{\mathcal{T}}^\chi$ of the elementary representation χ by

$$\left[\mathfrak{T}^{\chi}(g)\mathfrak{f}\right](k) = |a|^{h+c}\mathfrak{f}(k_g) \tag{2.24}$$

We can define a K-invariant scalar product in $C(K,\mathcal{V}^{\ell})$ (and, therefore, in each $\mathcal{C}_{\chi}$), by setting

$$(\mathfrak{f}_1,\mathfrak{f}_2) = \int_K dk < \mathfrak{f}_1(k), \mathfrak{f}_2(k)> \tag{2.25}$$

where < , > is the M-invariant scalar product on $\mathcal{V}^{\ell}$, and dk is the (normalized) Haar measure on K. The representation (2.24) is continuous with respect to the topology defined by this scalar product. There exists also another natural (Fréchet space) topology under which the space $C(K,\mathcal{V}^{\ell})$, regarded as a space of infinitely differentiable vector valued functions on the unit sphere (2.23), is complete. The representation $\mathcal{C}_{\chi}$ so defined is obviously unitary, when restricted to the compact subgroup K of G, and the space $C(K,\mathcal{V}^{\ell})$(or its Hilbert space completion) can be decomposed into a direct sum of (unitary) IR spaces $\mathcal{V}_{\tau}$ of K. It turns out that each IR τ of K appears at most once in a given elementary representation χ of G. This is a consequence of the following (Frobenius type)

Reciprocity Theorem (see [W3] Corollary 5.3.3.6 of Theorem 5.3.3.5). The elementary representation $\chi = [\ell, c]$ of G contains a given IR τ of K exactly as many times as τ contains the IR ℓ of M.

We have already recalled in Section 2. A that each IR ℓ of SO(2h)(=M) is contained at most once in a given IR τ of (SO(2h + 1)(=K). Hence we have the

Corollary 2.1. Each elementary representation $\chi = [\ell, c]$ of G = SO (2h + 1, 1) contains a given IR τ of K = SO(2h + 1) at most once. The representation τ is contained in χ if the IR ℓ of M is contained in τ.

The reduction of an elementary representation χ of G with respect to a chain of imbedded compact subgroups of the form $SO(2h + 1) = K \supset M^1 \supset \ldots \supset M^{2h-2}$ where

$$M^1 = M = SO(2h) \qquad M^j = SO(2h + 1-j)$$

provides a canonical basis in $C(K, \mathcal{V}^{\ell})$, because orthogonal groups are multiplicity free. If $\ell^{(1)}(= \ell), \ldots, \ell^{(2h-1)}$ are the multi-indices of the IR of M^1(=M), . . . , M^{2h-1} then the basic vectors of the canonical basis are given by $\Xi^{\tau}_{\ell^{(1)} \ldots \ell^{(2h-1)}}$. Such a basic vector transforms according to the IR $\ell^{(j)}$ of $M^{(j)}$ $j = 1, \ldots 2h-1$. The matrix elements of $\tilde{\mathfrak{T}}^{\chi}(g)$ in the canonical basis are called generalized spherical functions. Some of the functions will be displayed for the type I representations in Appendix A. 4.

2. D The noncompact picture: x-space realization

Most of our explicit calculations are performed in the x-space realization of the elementary representations based on the Bruhat decomposition (1. 13) (1. 15-18) of G.

We start again with the general covariant realization of Section 2. B. Using (1. 13) and (2. 14) we see that $\mathscr{f}(g)$ is completely fixed by its values

$$f(x) = \mathscr{f}(\tilde{n}_x) \tag{2. 26}$$

on the subgroup $\tilde{N} \approx R^{2h}$. The space of all (vector-valued) functions of this form, where $\mathscr{f}$ varies in $\mathscr{C}_\chi$ will be denoted by C_χ. Defining x_g, $a = a(g, x)$ and $m = m(g, x)$ from

$$g^{-1}\tilde{n}_x = \tilde{n}_{x_g} n^{-1} a^{-1} m^{-1} \tag{2. 27}$$

(cf. (1. 27)) we see that the tranformation law (2. 15) for $\mathscr{f}$ is equivalent to the following transformation law for f:

$$[T^\chi(g)f](x) = |a|^{-h-c} D^\ell(m) f(x_g) \tag{2. 28}$$

Defining $D^{\ell,c}(man) = |a|^{-h-c} D^\ell(m)$ and using (2. 27), we can rewrite (2. 28) as

$$[T^\chi(g)f](x) = D^{\ell,c}(\tilde{n}_x^{-1} g \tilde{n}_{x_g}) f(x_g) . \tag{2. 28'}$$

The expression for x_g for the various subgroups of G is obtained from (1. 15) (1. 17) (1. 28) (1. 29) by going to the inverse transformation. We shall write down the explicit form of T^χ for generic (Euclidean) conformal transformation in the special case of symmetric tensor representations (of type (2.9)) of M.

In this case the space C_χ consists of functions $f(x,\zeta)$ on $\mathbb{R}^{2h} \times \mathbb{K}_{2h}$ ($\mathbb{K}_{2h}$ being the complex isotropic cone(2. 11)), which have the following properties: f is a homogeneous polynomial of degree ℓ in ζ, infinitely differentiable in x and admits for $x \to \infty$ an asymptotic expansion of the form

$$f(x,\zeta) \underset{x\to\infty}{\approx} \frac{1}{(x^2)^{h+c}} \sum_{k=0}^{\infty} H_{k\ell}\,(Rx,\; r(x)\zeta). \tag{2. 29}$$

Here $H_{k\ell}$ is a homogeneous polynomial of degree k in the first argument and of degree ℓ in the second, R is the conformal inversion (1. 29) and $r(x) = m(R,x)$ is the O(2h) transformation, associated according to (2. 27) with the reflection R of the first 2h + 1 axes.

We have

$$R\tilde{n}_x = \tilde{n}_{Rx}\, n_x\, a\, r(x) \text{ with } |a| = |a(R,x)| = \frac{1}{x^2}, \tag{2. 30a}$$

and

$$r(x)^{\mu}{}_{\nu} = x^2\, \nabla^{\mu}\,(Rx)_{\nu} = 2\,\frac{x^{\mu} x_{\nu}}{x^2} - \delta^{\mu}{}_{\nu} \tag{2. 30b}$$

$$r^2(x) = r(Rx)r(x) = 1, \quad r(ax) = r(x) \quad (\text{for } a \in A), \quad r(x)x = x. \tag{2. 30c}$$

The space C_χ is complete with respect to a Fréchet space*) topology defined in terms of the Lie algebra generators by a countable set of (semi)norms (cf. [G2] where such a topology is introduced

---

*) For the functional analysis terminology, which is used without explanation, see e.g. [R1] ; Fréchet spaces are defined in Chapter V of that reference.

for the case of the ordinary Lorentz group; see also the remark at the end of this subsection.) The representation $T^{\chi}(g)$, $\chi = [\ell, c]$, is defined in the following way for the various subgroups of G_{ex}:

(a) Euclidean transformations: ($x' = (x'_1, \ldots, x'_{2h})$, $m \in O(2h)$)

$$[T^{\chi}(\tilde{n}_{x'} m) f](x, \mathfrak{z}) = f(m^{-1}(x-x'), m^{-1}\mathfrak{z}) \tag{2.31a}$$

(the formula is valid also for h = 1; in that case, for $\ell > 0$, the elementary representation becomes reducible when restricted to the connected subgroup G of G_{ex});

(b) dilatations:

$$[T^{\chi}(a) f(x, \mathfrak{z})] = \frac{1}{|a|^{h+c}} f\left(\frac{x}{|a|}, \mathfrak{z}\right), \quad |a| > 0 \tag{2.31b}$$

(c) conformal inversion:

$$[T^{\chi}(R)f](x, \mathfrak{z}) = \frac{1}{(x^2)^{h+c}} f(Rx, r(x)\mathfrak{z}). \tag{2.31c}$$

The special conformal transformations (1.28) are simply expressed in terms of the conformal inversion and a translation by b, so that it does not need to be considered separately.

We note that the assumed asymptotic behavior (2.29) of $f(x, \mathfrak{z})$ guarantees the smoothness of the right hand side of (2.31c) for $x \to 0$.

The equivalence between the compact and the noncompact pictures is displayed by the following relations:

$$f(x) = \frac{1}{(1+x^2)^{h+c}} \mathfrak{f}(k_x), \tag{2.32}$$

$$\mathfrak{f}(k) = \left(\frac{2}{1 + k^{2h+1}_{2h+1}}\right)^{h+c} D^{\ell}\left(m^{-1}(k)\right) f\left(x(k)\right), \tag{2.33}$$

where k_x is given by (1. 30a) while x(k) and m(k) are determined from (1. 32):

$$x(k)^{\mu} = -\frac{k^{\mu}_{2h+1}}{1+k^{2h+1}_{2h+1}}, \quad m(k)^{\mu}_{\nu} = k^{\mu}_{\nu} - \frac{k^{\mu}_{2h+1} k^{2h+1}_{\nu}}{1+k^{2h+1}_{2h+1}} . \tag{2.34}$$

The variables $\hat{\xi}^a = -k^a_{2h+1}$ $(a = 1, \ldots, 2h+1)$ (see (2. 23)) and x^{μ} are related to each other by a stereographic projection.

The above isomophism permits us to introduce a locally convex (Frechet space) topology on C_{χ} as the one induced by the natural topology on $C^{\infty}(S^{2h})$.

3. Further properties of the elementary representations

3.A Equivalence, irreducibility, completeness

We shall formulate here (without proofs) some known general properties of elementary representations. They can be, roughly, summarized as follows: Almost all elementary representations are irreducible. The exceptional--reducible--representations χ form a denumerable set with infinitesimal characters related to the finite dimensional representations. Any irreducible representation of G is equivalent to some subrepresentation of an elementary representation.

In order to state these results more precisely we shall need some of the general notions of the representation theory of locally compact groups (see Chapter 4 of ref. [W3]).

Let T and T' be two continuous representations of G acting in the Fréchet spaces $\mathfrak{X}$ and $\mathfrak{X}'$, respectively. We say, that the representations T and T' are partially equivalent if there exists a continuous linear map A: $\mathfrak{X} \to \mathfrak{X}'$ such that

$$AT(g) = T'(g)A \quad \text{for all } g \in G. \tag{3.1}$$

A map A with this property is called an intertwining operator for the representations T and T'. If A in (3.1) has a continuous inverse (that is, if A is bijective and bicontinuous), then the representations T and T' are said to be equivalent. Two Hilbert space representations T and T' are called unitarily equivalent if the bijection A is isometric (then if T is unitary, so is T'). If there exists an intertwining map A with a non-trivial kernel ker A (= the set of points of $\mathfrak{X}$ mapped onto the zero vector of $\mathfrak{X}'$) then ker A is an invariant subspace of $\mathfrak{X}$. More generally, if the representations T and T' are partially equivalent, but not equivalent, then at least one of them is reducible.

Most of the results concerning equivalence and irreducibility of elementary representations are stated for continuous representations on Banach spaces (Banach representations, for short). In order to comply with this general framework we will replace each of the Fréchet spaces C_χ of Sec. 2 by its Hilbert space completion $\mathcal{H}_\chi$ with respect to the scalar product (2. 25).

According to Corollary 2. 1 of the Reciprocity Theorem (Sec. 2B) each elementary representation χ of G (in either $\mathcal{H}_\chi$ or C_χ) is simply redicible (i. e., multiplicity free) with respect to the maximal compact subgroup K of G. Therefore the elementary representations of G pertain to the class of K-finite representations defined as follows. Let T be any continuous representation of K on $\mathfrak{X}$ (in particular, it can be a restriction of a representation of G on $\mathfrak{X}$). Let $\hat{K}$ be the set of all equivalence classes of finite dimensional IR's of K. For each $\tau \in \hat{K}$, let ξ_τ denote the character of τ $\left(\xi_\tau(k) = \mathrm{tr}\, \tau(k)\right)$, and $d(\tau)$ be its dimension. According to Schur's orthogonality relations the operator

$$\prod_\tau = d(\tau)\, T(\bar{\xi}_\tau) = d(\tau) \int \bar{\xi}_\tau(k)\, T(k) dk\,, \tag{3. 2}$$

where dk is the normalized Haar measure on K, is a (continuous) projection. A (continuous) representation T of G is said to be K-finite if the projection $\prod_\tau$ has a finite rank (i. e., if the space $\mathfrak{X}_\tau = \prod_\tau \mathfrak{X}$ is finite dimensional) for every $\tau \in \hat{K}$. In a simpler language a representation T of G is called K-finite if each irreducible representation of K appears in T with a finite (or zero) multiplicity.

If T is a K-finite Banach representation of G then the (dense) set of K-finite vectors

$$\mathfrak{X}_K = \sum_{\tau\in\hat{K}} \mathfrak{X}_\tau = \{f\in\mathfrak{X}; f = \sum_{i=1}^{n} f_i, f_i\in\mathfrak{X}_{\tau_i}, n < \infty\}$$

($\mathfrak{X}_\tau = \Pi_\tau \mathfrak{X}$ being the subspaces of $\mathfrak{X}$ introduced above) has the following remarkable properties: Each vector $f\in\mathfrak{X}_K$ is analytic. In other words, the mapping $G \ni g \mapsto T(g)$ defines an analytic function on the group G (considered as an analytic manifold) with values in the Banach space $\mathfrak{X}$ (see Lemma 4.5.5.1 of ref. [W3]). This fact allows one to define a representation T_K on $\mathfrak{X}_K$ of the Lie algebra $\mathfrak{g}$ (and thus of the universal enveloping algebra $\mathfrak{A}_{\mathbb{C}} = \mathfrak{A}(\mathfrak{g}_{\mathbb{C}})$ of its complexification $\mathfrak{g}_{\mathbb{C}}$).

Now we are prepared to complete our list of equivalence concepts with a more subtle notion, which will be used in the formulation of the subrepresentation theorem below.

The representations T_K and T'_K of $\mathfrak{A}_{\mathbb{C}}$ on $\mathfrak{X}_K$ and $\mathfrak{X}'_K$ are called algebraically equivalent, if there is a nonsingular operator $A: \mathfrak{X}_K \to \mathfrak{X}'_K$ such that $T'_K = A\, T_K\, A^{-1}$ (no continuity requirement being imposed on A). The K-finite Banach representations T and T' of G (on $\mathfrak{X}$ and $\mathfrak{X}'$, respectively) are said to be infinitesimally or Naimark equivalent if the corresponding representations T_K and T'_K are algebraically equivalent.

For infinite dimensional representations there are several nonequivalent notions of irreducibility. Every elementary representation of G on $\mathcal{H}_\chi$ is operator irreducible (in the sense of Schur's lemma): each bounded operator in $\mathcal{H}_\chi$, which commutes with the representation operators $T_\chi(g)$, is a multiple of the identity (this follows from the results of [K3, K4, Z4, Z5] ; the stateme is true for all elementary representations of $SO^{\uparrow}(2h+1, 1)$; for $G = Spin(2n,1)$ $(n = 1, 2, \ldots)$, it is only valid for $c \neq 0$). Nevertheless, as we shall see, some of the elementary representations are <u>topologically reducible</u> in the sense that there exist nontrivial (closed) invariant subspaces in some of the $\mathcal{H}_\chi$. We shall present a heuristic discussion of the question: which elementary representations should be expected to be reducible?

First of all, one would expect that representations, whose label $(\ell_1, \ldots \ell_{[h]}, -h-c)$ coincides with the highest weight of a finite dimensional representation of G might contain a finite dimensional invariant subspace. The type I representations satisfying this condition can be written in the form

$$\chi^-_{\ell\nu} = (0, \ldots 0, \ell, \ell+\nu-1) = [\ell, 1-h-\ell-\nu], \ell = 0, 1, \ldots \ \nu = 1, 2, \ldots \quad (3.3a)$$

It will be shown in Sec. 6A that these representations are indeed reducible. Other elementary representations which may be suspected of being reducible are those, obtained from $\chi^-_{\ell\nu}$ by a chain of intertwining maps. Since the Casimir operators of G are multiples of the identity for each elementary representation (this is true for any semi-simple Lie group since the Casimir invariants are polynomials of the labels c of the character of A) they have to have the same values for every pair of representations χ and χ' which can be related by (a chain of) intertwining maps. There are four type I representations (including (3.3a)) with this property; they are $\chi^-_{\ell\nu}$ given by (3.3a) and

$$\chi^+_{\ell\nu} = [\ell, h+\ell+\nu-1] \quad (3.3b)$$

$$\chi'^-_{\ell\nu} = [\ell+\nu, 1-h-\ell] \quad (3.3c)$$

$$\chi'^{+}_{\ell\nu} = [\ell+\nu,\ h+\ell-1]\ ,\ (\ell=0,1,\ \ldots;\ \nu=1,\ 2\ \ldots). \qquad (3.3d)$$

This is easy to verify for the second order Casimir operator $\mathcal{C}_2$. As we shall see in Sec. 6E below, the eigenvalues $\mathcal{C}_2(\chi)$ can be obtained from the expression (2.4) for the group SO(2h+2) via analytic continuation in c. They are:

$$\mathcal{C}_2(\chi) = \tfrac{1}{2} X_{AB} X^{BA} = \sum_{i=1}^{[h]} \ell_{[h]+1-i}(\ell_{[h]+1-i}+2h-2i) + c^2-h^2 \qquad (3.4a)$$

$$= \ell\ (\ell+2h-2) + (\nu+\ell-1)(\nu+\ell+2h-1) \text{ for the representations (3.3)} \qquad (3.4b).$$

If we go beyond the class of type I representations we will find exactly 2[h]+2 elementary representations with the same Casimir operators (or with the same infinitesimal characters in the mathematical terminology, cf.[Z4, Z5]). For example, for h = 2, these are the representations $\chi = [\ell_1, \ell_2; c]$ given by

$$\chi^{\pm}_{\ell\nu} = [\,0, \ell\,; \pm\,(\ell+\nu+1)] \qquad (3.5a)$$

$$\chi'^{\pm}_{\ell\nu} = [\,0, \ell+\nu; \pm\,(\ell+1)] \qquad (3.5b)$$

$$\chi''^{\pm}_{\ell\nu} = [\pm(\ell+1),\ \ell+\nu;\ 0]\,. \qquad (3.5c)$$

It appears that all these representations are indeed related by a chain of intertwining operators [Z5, G1]. All but the last pair of representations turn out to be reducible. (The representations (3.5c) belong to the principal series of unitary IR, studied in Sec. 3D below.)

All nonexceptional type I representations (different from (3.3)) are topologically irreducible. This is most easily demonstrated by infinitesimal methods (see Hirai, ref. [H3])). The irreducible components of elementary representations (including the irreducible elementary representations) are called subrepresentations.

The importance of the class of elementary representations stems from the fact that every IR of G (in a certain sense, to be specified) is equivalent to some subrepresentation of an elementary representation.

The following statement is a consequence of a recent result of Casselman [C1] and of the fundamental subquotient theorem of Harish-Chandra [H1, W3]. (In the context of elementary representations of $SO^{\uparrow}(N,1)$ a subquotient is either a subrepresentation or a factor representation in an elementary representation.)

Subrepresentation Theorem. A representation π of G (or $\mathfrak{g}$) which is Naimark equivalent to an algebraically irreducible K-finite representation of $\mathfrak{g}$ is also Naimark equivalent to some subrepresentation of an elementary representation of G. In particular, every unitary IR is equivalent to some such representation.

3.B Characters of elementary representations

Let $\chi = [\ell, c]$ be an elementary representation of G in the covariant realization of Sec. 2B. For infinitely differentiable functions $\varphi(g)$ of compact support on G, the operator

$$\mathcal{T}^{\chi}(\varphi) = \int \varphi(g)\mathcal{T}^{\chi}(g)\, dg \tag{3.6}$$

is known (see Theorem 8.7.4 of ref. [W2] to be trace class. Therefore, one can define its trace

$$\Theta_{\chi}(\varphi) = \mathrm{Tr}\,\mathcal{T}^{\chi}(\varphi); \tag{3.7a}$$

we shall also use the heuristic distribution theoretic notation

$$\Theta_{\chi}(g) = \mathrm{Tr}\,\mathcal{T}^{\chi}(g), \qquad \Theta_{\chi}(\varphi) = \int \Theta_{\chi}(g)\,\varphi(g)\, dg\,. \tag{3.7b}$$

Θ_{χ} is called the character of the elementary representation χ. We will write down an integral representation for Θ_{χ}. To do this we first introduce two auxiliary functions.

Let, as usual, $m \in M$, $a \in A$; define

$$F_{\varphi}(ma) = |a|^{h} \int_{K\times N} \varphi(kmnak^{-1})\, dk\, dn \tag{3.8}$$

and, for $k_1, k_2 \in K$.

$$F_{\varphi}^{\chi}(k_1, k_2) = \int |a|^{h-c} D^{\ell}(m)\, \varphi(k_1 m n a k_2^{-1})\, dm\, dn\, da \tag{3.9}$$

where $D^{\ell}(m)$ is the representation ℓ of $M = SO(2h)$, defined in Sec. 2A.

F_φ (ma) is a scalar function on M while $F_\varphi^\chi(k_1, k_2)$ is a matrix valued function on K x K. It follows from the definition that the functions F satisfy the covariance conditions

$$F_\varphi(m' m a m'^{-1}) = F_\varphi(m a) \qquad (m' \in M', \ m \in M, \ a \in A) \tag{3.10}$$

$$F_\varphi^\chi(k_1 m_1, k_2 m_2) = D^\ell(m_1)^{-1} F_\varphi^\chi(k_1, k_2) D^\ell(m_2). \tag{3.11}$$

The proof of (3.11) as well as the verification of (3.10) for $m' \in M$ is trivial. The only non-trivial point is the validity of (3.10) for $m' = w$ (where w is given by (1.10)) It is an immediate consequence of the following

Lemma 3.1. (cf. [W2] Lemma 7.7.11). Consider the coset space G/A with elements $\dot{g} = knA$ and measure $d\dot{g} = dk\,dn$. The function F_φ can be written in the form

$$F_\varphi(h) = \Delta(h) \int d\dot{g} \ \varphi(ghg^{-1}), \text{ for } h = ma \in MA, \tag{3.12}$$

where $\Delta(h)$ is independent of φ and invariant under the restricted Weyl group.

Proof: Using the Iwasawa decomposition (1.11) for g we see that ghg^{-1} only depends on the coset $\dot{g}$ (since A is in the center of $MA \ni h$) and that consequently the integral in (3.12) is well defined and equal to

$$\int dk\,dn \ \varphi(k n_b h n_b^{-1} k^{-1}) . \tag{3.13}$$

Assume for the moment that $a \neq 1$; then the map $\mu : N \to N$ defined by

$$\mu : n \to n' = m^{-1} n m a n^{-1} a^{-1}$$

is bijective. Indeed, noting that

$$m^{-1} n_b m = n_{m^{-1}b}, \quad a n_b^{-1} a^{-1} = n^{-1}_{b/|a|}, \tag{3.14}$$

we find that the Jacobian of the transformation μ is (the matrix under the sign of the determinant is $2h \times 2h$):

$$|\det(m^{-1} - |a|^{-1})| = |\det[\exp\{-(\varphi_1 X_{12} + \varphi_2 X_{34} + \ldots + \varphi_{[h]} X_{2[h]-1,\,2[h]})\} - |a|^{-1}]| \tag{3.15a}$$

$$= |1 - |a|^{-1}|^{2(h-[h])} \prod_{i=1}^{[h]} (1 + |a|^{-2} - 2|a|^{-1}\cos\varphi_i) \geq |1 - |a|^{-1}|^{2h} > 0 . \tag{3.15b}$$

(In writing down (3.15a) we have used the invariance of the determinant under similarity transformations.) Thus the integral (3.13) is equal to:

$$|\det(m^{-1} - |a|^{-1})|^{-1} \int dk\, dn'\, \varphi(k m n' a k^{-1}).$$

Then (3.12) coincides with (3.8), if we set

$$\Delta(h) = |a|^h |\det(m^{-1} - |a|^{-1})| = \left|\sqrt{|a|} - \frac{1}{\sqrt{|a|}}\right|^{2(h-[h])} \prod_{i=1}^{[h]} \left(|a| + \frac{1}{|a|} - 2\cos\varphi_i\right). \tag{3.16}$$

The result will remain true (by continuity) for $a=1$, for which $\Delta(h) = \Delta(m)$ may vanish. It implies, in particular, that

$$F_\varphi(m) = 0 = \Delta(m) \quad \text{for half-integer } h . \tag{3.17}$$

To complete the proof of the lemma we only have to note the w-invariance of $\Delta(h)$ (indeed, each factor in the right-hand side of (3.16) is invariant under the transformation $w: a \to a^{-1}$).

Now we are ready to state the following result of Harish-Chandra [H1].

Theorem 3.2. The characters $\Theta_\chi(\varphi)$ of the elementary representations are given by:

$$\Theta_\chi(\varphi) = \int_K \operatorname{tr} F_\varphi^\chi(k,k)\,dk = \int_{MA} |a|^{-\sigma} [\operatorname{tr} D^\ell(m)]\, F_\varphi(ma)\, dm\, da \tag{3.18}$$

where the trace tr in both cases is taken in the finite dimensional vector space $\mathcal{V}^\ell$.

The proof of this theorem (see Sec. 8.8.2 of ref. [W2]) is so simple and straightforward that we shall sketch it here.

Let $f \in \mathcal{C}_\chi$; then

$$[T^\chi(\varphi) f](k_1) = \int \varphi(g) f(g^{-1}k_1)\, dg = \int \varphi(g) D^\ell(m) f(g^{-1}k_1 m) dg$$

for any $m \in M$. Performing the change of variables $g^{-1}k_1 m \to g$ and using the invariance of dg and its expression (1.40) in terms of the Iwasawa factors, we obtain

$$[T^\chi(\varphi) f](k_1) = \int \varphi(k_1 m g^{-1}) D^\ell(m) f(g)\, dg$$

$$= \int \varphi(k_1 m a^{-1} n^{-1} k^{-1}) D^\ell(m) f(kna)\; dk\, dn\, da$$

Using again the covariance condition (2.14), performing the change of variables $n' = a^{-1} n^{-1} a$ $(|a|^{2h} dn = dn')$ $a' = a^{-1}$ and integrating the result with respect to $m \in M$ we find $(\int_M dm = 1)$,

$$[\mathcal{T}^{\chi}(\varphi) f](k_1) = \int F_{\varphi}^{\chi}(k_1, k) f(k)\, dk. \tag{3.19}$$

Thus, we presented $\mathcal{T}^{\chi}(\varphi)$ as a covariant integral operator with kernel (3.9). To compute the trace of such an operator one writes down its matrix elements in the canonical basis (defined at the end of Sec. 2. C). The result of this calculation (see [W2], p. 246) is

$$\mathrm{Tr}\, \mathcal{T}^{\chi}(\varphi) = \int \mathrm{tr}\, F_{\varphi}^{\chi}(k, k)\, dk .$$

This proves the first equation (3.18). The second one then follows from the definitions (3.8), (3.9).

<u>Corollary 3.3.</u> <u>Set for any</u> $\chi = [\ell, c]$

$$\tilde{\chi} = [\tilde{\ell}, -c] \tag{3.20}$$

<u>where</u> $\tilde{\ell}$ $(= \pi \ell)$ <u>is the representation of</u> M <u>obtained from</u> ℓ <u>by space reflection</u> <u>(see (2.8))</u>. <u>Then the elementary representations</u> χ <u>and</u> $\tilde{\chi}$ <u>have the same character:</u>

$$\Theta_{\chi}(\varphi) = \Theta_{\tilde{\chi}}(\varphi) . \tag{3.21}$$

<u>Proof:</u> if w is the non-trivial element (1.10) of the restricted Weyl group it acts on M as a space reflection, and therefore,

$$D^{\ell}(wmw^{-1}) = D^{\tilde{\ell}}(m) . \tag{3.22}$$

In order to obtain (3.21), we apply a Weyl transformation to (3.18) and use (3.22), the transformation law $waw^{-1} = a^{-1}$ for dilatations, and the w-invariance of F (ma).

3.C The spherical trace function. The character of a subquotient of an elementary representation

A useful expression for the character Θ, also applicable when U is a subrepresentation (or a factor) of an elementary representation is given in terms of the so-called spherical trace function defined below.

Let U be a subquotient of an elementary representation of G which contains the IR τ of K. Let further Π_τ be the projection operator on the corresponding subspace $\mathcal{V}_\tau$. The spherical trace function $t^U_\tau(g)$ is defined by:

$$t^U_\tau(g) = \mathrm{Tr}\,(\Pi_\tau\, U(g)\, \Pi_\tau) . \tag{3.23}$$

Proposition 3.4. Let U be equivalent to a subquotient of the elementary representation χ of $G \simeq SO^{\uparrow}(2h+1, 1)$ (or Spin (2h+1, 1)) which contains the representation τ of K; then $t^U_\tau(g) = t^\chi_\tau(g)$.

Proof: Assume that the representation space C_χ of an elementary representation χ contains an invariant subspace I_χ (possibly trivial). Then by Corollary 2.1 of the reciprocity theorem, if an IR τ of K is contained in χ, then it is contained either in the subrepresentation acting in I_χ or in the factor representation acting in C_χ/I_χ, but not in both. Thus $\Pi_\tau U(g)\Pi_\tau = \Pi_\tau T_\chi(g)\Pi_\tau$ and the proposition follows.

Corollary 3.5. Let U be any IR of $G_{(ex)}$ contained in a given elementary representation χ. Then the (distribution valued) character of U is given by

$$\Theta_U(g) = \sum_{\tau \in U} t^\chi_\tau(g) \tag{3.7c}$$

This formula is useful for the study of the discrete series (see Section 7A).

3.D The principal series of unitary representations

So far we have not discussed the question which of the elementary (sub)representations of G are unitary. This problem is closely related to the study of invariant bilinear forms on pairs of spaces C_χ. Here we shall start such a study by showing that the representations $\chi = [\ell, c]$ are unitary for c pure imaginary.

To do that we shall use the noncompact (x-space) realization of the elementary representations (see Section 2 D).

First of all we shall construct an invariant bilinear form on a pair of dual representations

$$\chi = [\ell, c], \quad \tilde{\chi} = [\ell, -c] \tag{3.24}$$

for arbitary (complex) c.

Proposition 3.6. If $f_1 \in C_{\tilde{\chi}}$, $f_2 \in C_\chi$, then the bilinear form

$$B(f_1, f_2) = \int <\bar{f}_1(x), f_2(x)> dx \tag{3.25}$$

(where $<f_1(x), f_2(x)>$ is the M-invariant scalar product in the finite dimensional vector space $\mathcal{V}^\ell$ -- cf. (2.25)) is invariant under the representation $T^{\tilde{\chi}} \otimes T^\chi$ of G.

Proof. Using the transformation law (2.28) for f_1 and f_2 we obtain

$$B(T^{\tilde{\chi}}(g)f_1, T^\chi(g)f_2) = \int |a(g,x)|^{-2h} <f_1(x_g), f_2(x_g)> dx, \tag{3.26}$$

where x_g and a(g, x) are defined from the Bruhat decomposition (2.27) of $g^{-1}\tilde{n}_x$. Noting that the Jacobian $\frac{Dx_g}{Dx}$ is nontrivial for dilatations and special conformal transofrmations of the type (1.28e) only, and using the relation

$$\frac{Dx_b}{Dx} = [1 - 2bx + b^2x^2]^{-2h} \quad \text{for } x_b = \frac{x - x^2 b}{1 - 2bx + b^2x^2},$$

we conclude that

$$|a(g,x)|^{-2h}\, dx = dx_g \,. \tag{3.27}$$

Inserting (3.27) in the right-hand side of (3.26) we reduce it to the original form (3.25). This completes the proof of the proposition.

Corollary 3.7. Let $f_1, f_2 \in C_\chi$ with $\chi = [\ell, c]$, c pure imaginary (c = $i\sigma$). Then the sesquilinear scalar product

$$(f_1, f_2) = \int \langle f_1(x),\, f_2(x)\rangle\, dx \tag{3.28}$$

is invariant under the action of the representation T_χ of G.

To reduce this statement to Proposition 3.6, it is sufficient to note that if $f(x) \in C_{[\ell, i\sigma]}$ (σ real), then the complex conjugate $\bar{f}(x) \in C_{[\ell, -i\sigma]}$.

The Hilbert space completion of C_χ with respect to the norm topology defined in terms of the scalar product (3.28) will be denoted by $\mathcal{H}_\chi$. We shall identify the representation T^χ (for $\chi = [\ell, i\sigma]$ with its extension to a unitary representation of G in $\mathcal{H}_\chi$. The family of unitary representations so constructed is called the (unitary) principal series. The irreducibility of these representations is guaranteed by a theorem by Hirai [H3] (see also [W3] vol. I p. 463). It, however, fails for the case of the two-fold covering Spin (2h + 1, 1) if 2h + 1 is even and c = o.

We mention finally that for unitary representations of the principal series the scalar product (3.28) coincides (up to a constant factor) with the scalar product (2.25) in the compact picture, so that our present notation $\mathcal{H}_\chi$ for the Hilbert space completion of C_χ does not conflict with the notation used in Section 3A.

3.E Infinitesimal generators and Casimir operators of the elementary representations

The emphasis in this work is on the global realization of the elementary representations. It is nevertheless useful to write down the expressions for the infinitesimal operators of these representations. We shall collect here the corresponding formulas in the noncompact picture. (They will be used, for example, in Section 6A, to establish intertwining properties of some differential operators for exceptional elementary representations of type (3.5).)

We shall restrict ourselves to type I representations (i.e., symmetric tensor inducing representations of M, --see (2.9)). In this case all infinitesimal generators appear as first order differential operators in x and $\mathfrak{z}$. We shall list them in the same order in which the corresponding global G transformations on $\tilde{N} \approx \mathbb{R}^{2h}$ were described in Section 1.C. (and the noncompact realization of the type I elementary representations was presented in Section 2 D). We shall deal throughout with the so-called mathematical generators (see Section 1 A and Table 1 of Section 1. B), which are antihermitian for the unitary representations of G.

(i) Translations. The generators T_μ, defined by

$T_\mu f(x) = \frac{\partial}{\partial x'^\mu} f(x-x')|_{x'=0}$, have the form

$$T_\mu = - \nabla_\mu \quad (= - \frac{\partial}{\partial x^\mu}) , \lambda = 1, \ldots , 2h. \tag{3.29}$$

(They are related to the Euclidean momentum operator P_μ by $P_\mu = iT_\mu$.)

(ii) Rotations (or M = SO(2h) transformations):

$$X_{\mu\nu} = - (x \wedge \nabla + \zeta \wedge \partial)_{\mu\nu} = x_\nu \nabla_\mu - x_\mu \nabla_\nu + \zeta_\nu \partial_\mu - \zeta_\mu \partial_\nu \tag{3.30}$$

We note that the "spinorial part" of $X_{\mu\nu}$,

$$s_{\mu\nu} = \zeta_\nu \partial_\mu - \zeta_\mu \partial_\nu , \quad \mu, \nu = 1, \ldots, 2h, \tag{3.31}$$

are interior differentiations on the cone $\mathbb{K}_n$ (2.11), since

$$[s_{\mu\nu} (\zeta^2 f(\zeta))]_{\zeta^2 = 0} = 0 \tag{3.32}$$

for any choice of $f(\zeta)$, and therefore, we do not need to introduce independent coordinates among the ζ's (cf. [33]).

(iii) Dilatations. Differentiating (2. 31b) with respect to $|a|$ and setting a = 1, we obtain

$$D\ (=X_{2h+1\ 0}) = -h - c - x\,\nabla. \qquad (3.33)$$

(iv) Special conformal transformations. They act non-trivially on both x and ζ:

$$C_\mu\ (=X_{\mu 0} - X_{\mu\, 2h+1}) = 2(h+c)x_\mu + 2x_\mu(x\nabla) - x^2\nabla_\mu + 2x^\lambda s_{\lambda\mu}, \qquad (3.34)$$

where $s_{\lambda\mu}$ is given by (3.31).

The Casimir operator

$$\mathcal{C}_2(\chi) = \tfrac{1}{2}X_{AB}X^{BA} = -\tfrac{1}{2}X^2_{\mu\nu} + D^2 + 2hD + C_\mu T_\mu \qquad (3.34)$$

turns out to be a constant. A straightforward computation of the right-hand side of (3.34) gives

$$\mathcal{C}_2(\chi) = \ell(\ell + 2h - 2) + c^2 - h^2. \qquad (3.35)$$

This expression coincides with the analytic continuation (3.4) of the eigenvalues (2.4) of the Casimir operator for the finite dimensional representations of $O^{\uparrow}(2h+1, 1)$ to arbitrary values of c. We note that, in agreement with (3.22), the Casimir operator (3.35) does not depend on the sign of c:

$$\mathcal{C}_2(\chi) = \mathcal{C}_2(\tilde{\chi}). \qquad (3.36)$$

II. INTERTWINING DISTRIBUTIONS AND THEIR FOURIER TRANSFORM

4. Intertwining operators: x-space realization

4.A Group theoretical definition of the intertwining operators

By a well known theorem (see [W3] vol. I, Corollary 4.5.8.3 to Theorem 4.5.8.1 and subsequent remark on p. 343) a character determines every K-finite unitary representation of G (in particular, every unitary IR) uniquely (up to equivalence). Since, according to (3.22), the unitary principal series representations $\chi = [\ell, c]$ and $\tilde{\chi} = [\tilde{\ell}, -c]$ have the same character, they must be equivalent. (We recall that according to (3.20) (2.8), if $\ell = (\ell_1 \ldots, \ell_{[h]})$, then $\tilde{\ell} = \left((-1)^{2(h-[h])}\ell_1, \ell_2, \ldots, \ell_{[h]}\right)$). Therefore, there should be an intertwining operator A_χ (see definition (3.1)) mapping $\mathcal{C}_{\tilde{\chi}}$ (and $\mathcal{H}_{\tilde{\chi}}$) onto $\mathcal{C}_\chi$ ($\mathcal{H}_\chi$) and commuting with the representation operators:

$$A_\chi \mathcal{T}^{\tilde{\chi}} = \mathcal{T}^{\chi} A_\chi \quad (\chi = [\ell, c], \ \tilde{\chi} = [\tilde{\ell}, -c]). \tag{4.1}$$

Let $f(g)$ belong to the space $\mathcal{C}_{\tilde{\chi}}$ of functions on G satisfying the covariance condition (2.14) with χ replaced by $\tilde{\chi}$. Then the intertwining operator A_χ can be defined by

$$[A_\chi \mathfrak{f}](g) = \int_{\widetilde{N}} \mathfrak{f}(gw\tilde{n}_x)dx = \int_N \mathfrak{f}(gn_b w)db, \tag{4.2}$$

where w is again the non-trivial element (1.10) of the restricted Weyl group. To verify this statement we have first to show that $A_\chi \mathfrak{f}$ satisfies the covariance properties for an element of $\mathcal{C}_\chi$; indeed,

$$(A_\chi \mathfrak{f})(gman) = \int_{\widetilde{N}} \mathfrak{f}(gman_b w\tilde{n}_{x'})\, dx' =$$

$$= \int_{\widetilde{N}} \mathfrak{f}(gwm^w a^{-1}\tilde{n}_{b'+x'})\, dx' =$$

$$= |a|^{2h} \int_{\widetilde{N}} \mathfrak{f}(gw\tilde{n}_x m^w a^{-1})\, dx = |a|^{h+c} D^{\tilde{\ell}^{-1}}(m^w) \int_{\widetilde{N}} \mathfrak{f}(gw\tilde{n}_x)\, dx =$$

$$= |a|^{h+c} D^{\ell}(m^{-1})\, (A_\chi \mathfrak{f})(g). \tag{4.3}$$

In deriving (4.3) we used the following (notation and) identities:

$$w^{-1} n_b w = w n_b w^{-1} = \tilde{n}_{b'}, \quad b' = (b_1, \ldots, b_{2h-1}, -b_{2h}), \tag{4.4a}$$

$$w^{-1} a w = w a w^{-1} = a^{-1}, \tag{4.4b}$$

$$w^{-1} m w = w m w^{-1} \equiv m^w, \quad D^{\tilde{\ell}}(m^w) = D^{\ell}(m), \tag{4.4c}$$

$$a^{-1}\tilde{n}_x a = \tilde{n}_{|a|^{-1}x} \tag{4.4d}$$

(cf. (1.10) and Section 3.B). The covariance property (4.1) is verified by a straightforward application of (2.15).

Using the Iwasawa decomposition (1.31) of $\tilde{n}_x$ and the covariance property (2.14) for $f \in \mathcal{C}_{\tilde{\chi}}$, we can rewrite (4.2) in the form

$$(A_\chi f)(g) = \int_{\tilde{N}} f(gwk_x) \frac{dx}{(1+x^2)^{h-c}} . \tag{4.5}$$

Since k_x varies in the compact manifold K we see that:

A_χ is well defined and analytic in c for Re $c < 0$; (4.6a)

similarly we deduce that:

$A_{\tilde{\chi}} : \mathcal{C}_\chi \to \mathcal{C}_{\tilde{\chi}}$ is well defined and analytic for Re $c > 0$. (4.6b)

Both A_χ and $A_{\tilde{\chi}}$ can be extended to meromorphic functions in the entire complex plane c(see [K3,K4,S1]).

From now on we shall consider the special case, when the representation ℓ of M is equivalent to its mirror image (see Section 2A) and can therefore be extended to a representation of $G_{ex} = O^{\uparrow}(2h+1, 1)$, including the space reflection

$$I_s(\underline{x}, x_{2h}) = (-\underline{x}, x_{2h}) \tag{4.7a}$$

and the (Euclidean) "time inversion"

$$\Theta(\underline{x}, x_{2h}) = (\underline{x}, -x_{2h}), \quad \underline{x} = (x_1, \ldots, x_{2h-1}). \tag{4.7b}$$

We note that I_s is either a proper rotation (for 2h odd) or related to Θ by the proper rotation $I_s\Theta$ (for 2h even).

Using (2.8) and the fact that the transformation $I_s\Theta$: $x \to -x$ commutes with $M = SO(2h)$ we find

$$D^{\tilde{\ell}}(m) \equiv D^{\ell}(m^{w}) = D^{\ell}(\Theta m \Theta^{-1}) = D^{\ell}(I_s m I_s^{-1}) = D^{\ell}(I_s) D^{\ell}(m) D^{\ell}(I_s)^{-1}. \quad (4.8)$$

Hence, we can define an equivalence map $\mathcal{T}(I_s)$: $\mathcal{C}_{[\tilde{\ell}, -c]} \to \mathcal{C}_{[\ell, -c]}$ by

$$[\mathcal{T}(I_s) f](g) = D^{\ell}(I_s) f(g). \quad (4.9)$$

(Note that we use distinct notation ℓ and $\tilde{\ell}$ for the mirror image representations even if they are equivalent, since the matrices $D^{\tilde{\ell}}$ and D^{ℓ}, appearing in the equivalence relation (4.8), are different except for the trivial representation $\ell = 0$.)

We define the normalized intertwining operators G_χ: $\mathcal{C}_{\tilde{\chi}} = \mathcal{C}_{[\ell, -c]} \to \mathcal{C}_{[\ell, c]} = \mathcal{C}_\chi$ and $G_{\tilde{\chi}}$: $\mathcal{C}_\chi \to \mathcal{C}_{\tilde{\chi}}$ by

$$G_\chi = \gamma_\chi A_\chi \mathcal{T}(I_s), \quad G_{\tilde{\chi}} = \gamma_{\tilde{\chi}} \mathcal{T}(I_s) A_{\tilde{\chi}} \quad (4.10)$$

where γ_χ is a normalization factor to be determined in such a way that the operators G_χ satisfy the normalization condition

$$G_\chi G_{\tilde{\chi}} = 1 = G_{\tilde{\chi}} G_\chi \quad (4.11)$$

for c pure imaginary. Condition (4.11) can indeed be satisfied, since $G_\chi G_{\tilde{\chi}}$ is a unitary map from $\mathcal{H}_\chi$ onto itself which commutes with all $T_\chi(g)$ $(g \in G_{ex})$. Schur's lemma (asserting that $G_\chi G_{\tilde{\chi}}$ is a multiple of the identity in $\mathcal{H}_\chi$) can be applied--according to Section 3.A.--to all elementary representations.

The operator G_χ (considered as an integral operator in the noncompact picture) appears as analytic continuation (in c) of the Euclidean 2-point function in a conformal covariant quantum field theory [M6, M7, T5, M2]. (This accounts, in particular, for our choice of notation.)

4.B The intertwining distribution in the noncompact picture

Now we shall find the action of the operator G_χ on functions $f(x) \in C_{\tilde{\chi}}$ (related to $\mathfrak{f}(g)$ by (2.26)). According to (4.10) (4.2) we have

$$[G_\chi f](x_1) = \gamma_\chi \int D^\ell(I_s)\, \mathfrak{f}(\tilde{n}_{x_1} w \tilde{n}_x)\, dx =$$

$$= \gamma_\chi \int D^\ell(I_s m(x,w)^{-1}) f(x+x')\,(x^2)^{c-h} dx \,, \tag{4.12}$$

In deriving the last equality we have used the relation

$$w\tilde{n}_x = \tilde{n}_{x'}\, n_{I_s x}\, a(x,w)\, m(x,w), \qquad |a(x,w)| = \frac{1}{x^2} \tag{4.13a}$$

where

$$x' = I_s Rx = \left(\frac{\underline{x}}{x^2}, \frac{-x_{2h}}{x^2} \right), \quad m = m(x,w) = I_s\, r(x), \quad m^{-1} = r(x)\, I_s \tag{4.13b}$$

and $r(x)$ is given by (2.30). Performing two consecutive changes of variables $x \to x'$: $x = I_s Rx'$ $\left(x^2 = \frac{1}{x'^2},\ dx = (x'^2)^{-2h} dx'\right)$ and $x' \to x_2$: $x' = x_2 - x_1$ in (4.12) and using the identity

$$I_s\, r\,(I_s Rx')I_s = r(x') \tag{4.14}$$

(which is implied by (2.30)), we obtain

$$(G_\chi f)(x_1) = \int G_\chi (x_1 - x_2)\, f(x_2)\, dx_2, \tag{4.15}$$

where $G_\chi (x_{12})$ is the (matrix) kernel

$$G_\chi (x_{12}) = \frac{\gamma_\chi}{(x_{12}^2)^{h+c}}\, D^\ell (r(x_{12})), \quad x_{12} = x_1 - x_2 . \tag{4.16}$$

(For fixed x_{12}, G_χ is a (matrix) operator in the finite dimensional representation space $\mathcal{V}^{\ell}$.)

In the special case of type I (symmetric tensor) representations of O(2h) (see (2.9)) we can replace D^ℓ by the corresponding generating homogeneous polynomial

$$\mathfrak{z}_1^{\otimes \ell} D^\ell (r(x))\, \mathfrak{z}_2^{\otimes \ell} = (\mathfrak{z}_1\, r(x)\, \mathfrak{z}_2)^\ell, \text{ for } \mathfrak{z}_1, \mathfrak{z}_2 \in \mathbb{K}_{2h} \tag{4.17}$$

(it suffices to assume that at least one of the variables $\mathfrak{z}_1$ or $\mathfrak{z}_2$ satisfies the light-cone condition $\mathfrak{z}^2 = 0$).

Expressing γ_χ in terms of another normalization constant $n(\chi)$,

$$\gamma_\chi = \frac{n(\chi)(-1)^\ell}{(2\pi)^h} 2^{h+c}, \qquad (4.18)$$

in order to conform to the notation of refs. [4, 5], we obtain

$$G_\chi(x;\mathfrak{z},\zeta) = \frac{n(\chi)}{(2\pi)^h}\left(\frac{2}{x^2}\right)^{h+c}(-\mathfrak{z}r(x)\zeta)^\ell \quad (\mathfrak{z}\in K_{2h},\ \zeta\in\mathbb{C}^{2h}) \qquad (4.19)$$

The expression for the Green function (without using the $\mathfrak{z}$, ζ formalism) was first obtained from $O^\uparrow(2h+1, 1)$ invariance consideration (see, e.g., [F4, M10, P3]). The derivation of (4.16) presented here which follows the mathematicians' work [K3, K4, W2], is due to Koller [K5]. It makes transparent the role of the Weyl symmetry w and the conformal inversion R.

5. Momentum space expansion of the intertwining distribution and positivity

5.A Fourier transform of $G_\chi(x;\mathfrak{z}_1,\mathfrak{z}_2)$

The intertiwining kernels are used, among other things, to construct scalar products for the complementary (and some exceptional "discrete") series of unitary representations of G.

It is, therefore, important to investigate their positivity properties. It turns out that this is most conveniently done by carrying out a harmonic analysis of G_χ on the parabolic subgroup $\tilde{N}AM$.

The group $\tilde{N}AM$ is a semi-direct product of an abelian subgroup $\tilde{N}$ and a reductive group AM which is the direct product of a compact semisimple part M and of a one-dimensional dilatation subgroup A. The harmonic analysis on such groups is elementary and proceeds in two steps. First, one performs the (ordinary) Fourier transform on $\tilde{N} \simeq \mathbb{R}^{2h}$, defining

$$G_\chi(p;\mathfrak{z}_1,\mathfrak{z}_2) = \int G_\chi(x;\mathfrak{z}_1,\mathfrak{z}_2)e^{-ipx}\,dx. \tag{5.1}$$

Using the integration formula

$$\frac{\Gamma(h+c)}{(2\pi)^h}\int\left(\frac{2}{x^2}\right)^{h+c} e^{-ipx}dx = \frac{1}{(2\pi)^h}\int_0^\infty d\alpha\,\alpha^{h+c-1}\int dx\, e^{-\frac{1}{2}\alpha x^2 -ipx}$$

$$= \int_0^\infty d\alpha\,\alpha^{c-1}\exp\left(-\frac{1}{2\alpha}p^2\right) = \Gamma(-c)\left(\tfrac{1}{2}p^2\right)^c \tag{5.2}$$

valid for $-1 < \operatorname{Re} c < 0$, and extending it by analytic continuation as a distribution theoretic identity to all noninteger c, we obtain

$$G_\chi(p;\mathfrak{z}_1,\mathfrak{z}_2) = \frac{n(\chi)}{(2\pi)^h}\sum_{k=0}^{\ell}\binom{\ell}{k}(\mathfrak{z}_1\partial_p)^k(\mathfrak{z}_2\partial_p)^k(\mathfrak{z}_1\mathfrak{z}_2)^{\ell-k}\int\left(\frac{2}{x^2}\right)^{h+c+k}e^{-ipx}dx =$$

$$= \frac{n(\chi)(\frac{1}{2}p^2)^c}{\Gamma(c+h+\ell)\Gamma(c+h-1)} \sum_{k=0}^{\ell} \binom{\ell}{k} \Gamma(\ell-k-c)\Gamma(h+k+c-1) \left[\frac{(p\zeta_1)(p\zeta_2)}{\frac{1}{2}p^2}\right]^{\ell-k} (\zeta_1\zeta_2)^k =$$

$$= (-1)^{\ell}\ell! \frac{n(\chi)\Gamma(-c)}{\Gamma(c+h+\ell)} \left[\frac{(p\zeta_1)(p\zeta_2)}{\frac{1}{2}p^2}\right]^{\ell} (\tfrac{1}{2}p^2)^c \; P_{\ell}^{(c-\ell,\; h-2)}(\omega) , \tag{5.3}$$

where

$$\omega = \cos\theta = 1 - \frac{p^2(\zeta_1\zeta_2)}{(p\zeta_1)(p\zeta_2)} \tag{5.4}$$

[θ being the angle between the (2h-1) vectors $\underline{\zeta}_1$ and $\underline{\zeta}_2$ in the "rest frame"

$$p = (\underline{0}, |p|) \qquad |p| = (p^2)^{\frac{1}{2}} \tag{5.5}$$

of p]. Here we have used the following expansion formula for the Jacobi polynomials (see Eq. 8.962.1 of ref. [G7]):

$$P_{\ell}^{(c-\ell,\; h-2)}(\omega) = \frac{(-1)^{\ell}}{\ell!} \sum_{k=0}^{\ell} \binom{\ell}{n} \frac{\Gamma(\ell-1-c)\Gamma h+k+c-1)}{\Gamma(-c)\Gamma(c+h-1)} \left(\frac{1-\omega}{2}\right)^k . \tag{5.6}$$

For real c, if $f(x) \in C_{\chi}$, then also $\overline{f}(x) \in C_{\chi}$ (and we could have considered the restriction of C_{χ} to real f as a real T^{χ} covariant vector space if we wished to). For such c, G_{χ} defines a hermitian sesquilinear form on $C_{\tilde{\chi}} \times C_{\tilde{\chi}}$:

$$(f_1, G_\chi f_2) = \int\int <f_1(x_1), G_\chi(x_{12}) f_2(x_2)> dx_1 \, dx_2 =$$

$$= \int <f_1(p), G_\chi(p) f_2(p)> \frac{dp}{(2\pi)^{2h}} . \qquad (5.7)$$

(Following the physicist tradition we distinguish between functions and their Fourier transforms only by their arguments.) Thus, we can expect to find complementary series of unitary representations of $G_{(ex)}$ for some real values of c only.

5.B Harmonic expansion of $G_\chi(p)$

Since the Fourier transform (5.3) of $G_\chi(x;\mathfrak{z}_1,\mathfrak{z}_2)$ is automatically homogeneous in p (and therefore irreducible with respect to A) it remains as a second step to carry out the harmonic analysis with respect to the stability group

$$U = SO(2h-1)_p \qquad (5.8)$$

of p. In order to perform this step we shall use the frame (5.5) (which is possible by O(2h) covariance) and will set

$$\mathfrak{z}_j = (\underline{z}_j, i), \quad \underline{z}_j \text{ - real}, \quad \underline{z}_j^2 = 1, \quad j = 1, 2; \quad \underline{z}_1 \underline{z}_2 = \cos\theta . \qquad (5.9)$$

(That is no restriction on generality, because G_χ is a homogeneous polynomial in $\mathfrak{z}_1$ and $\mathfrak{z}_2$.)

We consider the Jacobi polynomial $P_\ell^{(c-\ell,\ h-2)}(\underline{z}_1\underline{z}_2)$ as a function of $\underline{z}_2$ for fixed $\underline{z}_1$) and carry out its harmonic analysis. Since it depends only on the scalar product $\underline{z}_1\ \underline{z}_2$, it is invariant under the stability subgroup $SO(2h-2)_{\underline{z}_1}$ of $\underline{z}_1$, so that one is lead to the decomposition of $P_\ell^{(c-\ell,\ h-2)}(\underline{z}_1\underline{z}_2)$ in zonal spherical functions $C_s^{h-\frac{3}{2}}(\underline{z}_1\ \underline{z}_2)$ of SO(2h-1) (defined in Appendix A. 2):

$$P_\ell^{(c-\ell,\ h-2)}(\omega) = (h-1)_\ell \sum_{s=0}^{\ell} \frac{2h+2s-3}{(\ell-s)!} \frac{(c+h-1)_s}{(2h-3)_{\ell+s+1}} (c-h-\ell+2)_{\ell-s}\, C_s^{h-\frac{3}{2}}(\omega), \qquad (5.10)$$

where we use the notation

$$(\alpha)_k = \frac{\Gamma(\alpha+k)}{\Gamma(\alpha)} = \alpha\ (\alpha+1)\ \ldots(\alpha+k-1). \qquad (5.11)$$

(Let us caution the reader that in deriving (5.10) we had to correct some misprints in Eq. 7.391.10 of ref. [G7].)

By construction, the expansion (5.10) corresponds to the decomposition of the Green function $G_\chi(p)$ (regarded as an operator in the finite dimensional space $\mathcal{V}^\ell = \mathcal{V}^\ell_{(2h)}$) into $SO(2h-1)_p$ projection operators $\Pi^{\ell s}(p)$, which map $\mathcal{V}^\ell_{(2h)}$ onto the subspace $\mathcal{V}^s_{(2h-1)}$ of SO(2h -1) symmetric traceless tensors of rank $s \leq \ell$. [The easiest way to see that is to note that due to Eqs. (A. 18) and (A. 19) of Appendix A the zonal spherical function $C_s^{h-\frac{2}{3}}$ belongs to the eigensubspace of the Casimir operator

$\Omega\,[2h-1]$ of SO(2h-1), corresponding to eigenvalue s(s+2h-3).] These

projection operators have the following characteristic properties:

$$\Pi^{\ell s}(p)\,\Pi^{\ell s'}(p) = \delta_{ss'}\,\Pi^{\ell s}(p)\,, \tag{5.12}$$

$$\sum_{s=0}^{\ell} \Pi^{\ell s}(p) = 1, \tag{5.13}$$

$$p^{\otimes \ell-s+1}\,\Pi^{\ell s}(p) \equiv p^{\mu_s}\ldots p^{\mu_\ell}\;\Pi^{\ell s}_{\mu_1\ldots\mu_s\ldots\mu_\ell}(p)_{\nu_1\ldots\nu_\ell} = 0 =$$

$$= \Pi^{\ell s}(p)p^{\otimes \ell-s+1}, \quad \text{for } 0 < s \le \ell. \tag{5.14}$$

The completeness relation (5.13) follows from the remark at the end of Section 2. A, which says that type I representations of SO(2h) contain only type I representations of SO(2h-1). Eq. (5.14) follows from the argument given below.

Consider the operator Q: $\mathcal{V}^{\ell}_{(2h)} \to \mathcal{V}^{\ell-1}_{(2h)}$, defined by $f = (f_{\mu_1\ldots\mu_\ell}) \to Qf = (p^{\mu_\ell} f_{\mu_1\ldots\mu_{\ell-1}\mu_\ell})$. It is obviously $U \simeq$ SO(2h-1)-covariant and satisfies

$$QD^{\ell}(u) = D^{(\ell-1)}(u)Q, \qquad Q\,\Pi^{\ell s} = \Pi^{\ell-1s}Q.$$

Applying the second equality $\ell - s + 1$ times we obtain

$$Q^{\ell-s+1}\Pi^{\ell s} = \Pi^{s-1s}Q^{\ell-s+1} = 0$$

(since $\Pi^{\ell s} = 0$ for $\ell < s$). This proves (5.14).

In terms of the generating functions

$$\Pi^{\ell s}(p;\mathfrak{z}_1,\mathfrak{z}_2) = \mathfrak{z}_1^{\mu_1}\ldots\mathfrak{z}_1^{\mu_\ell}\,\Pi^{\ell s}_{\mu_1\ldots\mu_\ell\nu_1\ldots\nu_\ell}(p)\mathfrak{z}_2^{\nu_1}\ldots\mathfrak{z}_2^{\nu_\ell} \qquad (5.15)$$

the completeness relation (5.13) assumes the form

$$\sum_{s=0}^{\ell}\Pi^{\ell s}(p;\mathfrak{z}_1,\mathfrak{z}_2) = (\mathfrak{z}_1\mathfrak{z}_2)^\ell\left[= (\omega-1)^\ell\right]; \qquad (5.16)$$

the last equality (in brackets) hold for the special choice (5.5) (5.9) of the vectors $p, \mathfrak{z}_1$, and $\mathfrak{z}_2$. It follows from the preceding discussion that

$$\Pi^{\ell s}(p;\mathfrak{z}_1\mathfrak{z}_2) = A_{\ell s}(-1)^s\left[\frac{(p\mathfrak{z}_1)(p\mathfrak{z}_2)}{\frac{1}{2}p^2}\right]^\ell C_s^{h-\frac{3}{2}}(\omega). \qquad (5.17)$$

Here $A_{\ell s}$ is a normalization constant determined (in principle) from (5.12). In order to evaluate it, we insert (5.17) in the completeness relation (5.16); the result is

$$\sum_{s=0}^{\ell}(-1)^s A_{\ell s}C_s^{h-\frac{3}{2}}(\omega) = \left(\frac{1-\omega}{2}\right)^\ell. \qquad (5.18)$$

Using the orthogonality property for Gegenbauer polynomials and their normalization (see (A.20)) as well as the integration formula

$$(-1)^s\, 2^{-\ell} \int_{-1}^{1} (1-\omega)^{\ell+h-2} (1+\omega)^{h-2}\, C_s^{h-\frac{3}{2}}(\omega)\, d\omega = \binom{\ell}{s} \Gamma(h-1)\Gamma(\ell+h-1) \frac{2^{2h-3}(2h+s-4)!}{(2h-4)!\,(2h+\ell+s-3)!}$$

(see [G7] Eq. 7 311. 3), we obtain

$$A_{\ell s} = (2h + 2s - 3) \frac{\ell!}{(\ell - s)!} \quad \frac{(h-1)_\ell}{(2h-3)_{\ell+s+1}} . \tag{5.19}$$

Noting that

$$\lim_{\nu \to 0} \frac{1}{2\nu} C_s^{\nu}(\cos\theta) = \frac{1}{s} \cos s\theta$$

(see [G7] Eq. 8. 934. 4) we see that in the special case when 2h = 3 and Eq. (5.19) becomes meaningless, the projection operators (5.17) can still be defined and are equal to

$$\Pi^{\ell s}_{(2h=3)}(p, \mathfrak{z}_1, \mathfrak{z}_2) = (-1)^s \frac{2\,\ell!\,(\frac{1}{2})_\ell}{(\ell-s)!\,(\ell+s)!} \left[\frac{(p\mathfrak{z}_1)(p\mathfrak{z}_2)}{\frac{1}{2}p^2} \right]^\ell \cos s\theta, \tag{5.20}$$

where $\cos\theta$ $(=\omega)$ is given by (5. 4). In this special case $\Pi^{\ell s}$ for $s > 0$ is a projection on a two-dimensional space which is reducible with respect to the proper rotation group SO(2), but becomes irreducible, if we consider the group O(2) including reflections. The decomposition of $\Pi^{\ell s}$ into one dimensional SO(2)-irreducible projectors corresponds to the decomposition of $\cos s\theta$ into $e^{\pm is\theta}$. (We note that in this case $\sin\theta = (p\mathfrak{z}_1 p\mathfrak{z}_2)^{-1}\sqrt{p^2}\,(p\mathfrak{z}_1 \Lambda \mathfrak{z}_2)$.) The label s of SO(2)-irreducible representations will then vary between $-\ell$ and ℓ.

The preceding discussion does not apply to the case $h = 1$ (corresponding to the ordinary Lorentz group $O^{\uparrow}(3,1)$) which is exceptional from our present point of view. That case is treated in Appendix B, where it is shown, in particular, that the operators $\Pi^{\ell s}$ vanish for $s > 1$ $(h=1)$.

In order to be able to write down in a simple way the scalar product of two symmetric tensors in terms of the corresponding homogeneous polynomials f, it is convenient to extend $f(\mathfrak{z})$ to a homogeneous harmonic polynomial $F(\zeta)$ $\left(=f_H(\zeta)\right)$ $\zeta \in \mathbb{C}^{2h}$ (see Appendix A). The factorization property of the projection operator $\Pi^{\ell s}(p;\mathfrak{z}_1,\mathfrak{z}_2)$, exhibited in (5.17), persists for its harmonic extension. It is related to the decomposition of an arbitrary SO(2h)-transformation into a rotation in the plane (2h-1, 2h) sandwiched between two SO(2h-1) transformations (which leave the axis 2h invariant). Its group theoretical derivation (given in Appendix A.4) also uses the factorization and orthogonality properties of generalized spherical functions. The results for the harmonic extension of (5.17) is

$$\Pi^{\ell s}(p;\zeta_1\zeta_2) = b_{\ell s}\, L_{\ell s}(p,\zeta_1)\, \Pi^{ss}(p;\zeta_1,\zeta_2)\, L_{\ell s}(p,\zeta_2) \tag{5.21a}$$

where ζ_1, ζ_2 are arbitrary complex 2h-vectors $(\zeta_1,\zeta_2 \in \mathbb{C}^{2h})$.

$$2^{\ell-s}\binom{h+\ell-2}{\ell-s} L_{\ell s}(p,\zeta) = (\zeta^2)^{\frac{\ell-s}{2}}\, C^{h+s-1}_{\ell-s}(\hat{p}\hat{\zeta}) \qquad \hat{p} = (p^2)^{-\frac{1}{2}}p\,, \quad \hat{\zeta} = (\zeta^2)^{-\frac{1}{2}}\zeta\,, \tag{5.21b}$$

$$\left[L_{\ell s}(p,\mathfrak{z}) = (\hat{p}\mathfrak{z})^{\ell-s} \quad \text{for } \mathfrak{z}^2 = 0\right]$$

$$\Pi^{ss}(p;\zeta_1,\zeta_2) = (-2)^s \frac{s!\,(h-1)_s}{(2h-3)_{2s}} (\pi_{11}\pi_{22})^{s/2} C_s^{h-\frac{3}{2}}(\omega), \tag{5.21c}$$

$$\omega = \frac{-\pi_{12}}{\sqrt{\pi_{11}\pi_{22}}},$$

$$b_{\ell s} = 2^{\ell-s} \frac{(s+h-1)_{\ell-s}}{(2s+2h-2)_{\ell-s}} \binom{\ell}{s}, \tag{5.21d}$$

and π_{ij} are the elementary "projection variables"

$$\pi_{ij} = \Pi^{11}(p;\zeta_i,\zeta_j) = \zeta_i\zeta_j - \frac{(p\zeta_i)(p\zeta_j)}{p^2}, \quad i,j = 1,2. \tag{5.22}$$

The simplest way to prove (5.21) consists in verifying that the right hand side is harmonic in both ζ_1 and ζ_2, and goes into (5.17) for (5.19) to $\zeta_j = \mathfrak{z}_j$ ($\mathfrak{z}_j^2 = 0$). We shall apply this factorized form for the projection operators to the derivation of some differential identities among the intertwining operators at exceptional integer points in Section 6. D.

5.C Normalization and positivity for non-exceptional representations. Complementary series of unitary IR's

Putting together Eqs. (5.3)(5.10) (5.17) and (5.19), we obtain the following harmonic ($SO(2h-1)_p$) expansion for the momentum space intertwining kernel:

$$G_\chi(p) = \frac{n(\chi)\Gamma(-c)(\frac{1}{2}p^2)^c}{(h+\ell+c-1)\Gamma(h+c-1)} \sum_{s=0}^{\ell} \kappa_{\ell s}(c)\, \Pi^{\ell s}(p), \tag{5.23a}$$

where

$$\kappa_{\ell s}(c) = (-1)^{\ell-s} \frac{\Gamma(h+c+s-1)\Gamma(c+2-h-s)}{\Gamma(h+c+\ell-1)\Gamma(c+2-h-\ell)} = \frac{(h+s-c-1)_{\ell-s}}{(h+s+c-1)_{\ell-s}} . \tag{5.23b}$$

The normalization condition (4.11) does not fix the factor $n(\chi)$ uniquely and there seems to be no universal choice of $n(\chi)$ equally suited for all purposes. Knapp and Stein [K3] [Section 13 (Lemma 36)] require (in addition to (4.11)) that $n(\chi)$ is a meromorphic function of c with all its zeros in the (closed) right half plane and all its poles in the left half plane and that it is real for real c. The simplest choice of this type which satisfies the above conditions for both integer and half integer h, and has only simple poles and zeros, is given by

$$(n(\chi) =)\ n_+(\chi) = (h+\ell+c-1)\frac{\Gamma(h+c-1)}{\Gamma(-c)} . \tag{5.24}$$

The normalization property (4.11) follows from the identity

$$\kappa_{\ell s}(c)\kappa_{\ell s}(-c) = 1$$

for the coefficients (5.23b).

We shall also use three other choices of $n(\chi)$ [all compatible with (4.11)] each appropriate for a different kind of problem: The normalization

$$n_o(\chi) = \frac{\Gamma(h+\ell+c)\Gamma(h-c-1)}{\Gamma(h+\ell-c-1)\Gamma(-c)}, \qquad (5.25)$$

adopted in refs. [M2,D4] is particularly convenient in writing down the differential identities between hermitian forms for exceptional representations (see Section 6. D below). The corresponding intertwining operator becomes infinite for the exceptional representations $\chi^-_{\ell\nu}$ (3.3a) and $\chi'^+_{\ell\nu}$ (3.3d).

In the derivation of operator product expansion in the framework of quantum field theory pursued in [M2,D4] (see also Chapter V), ere one is concerned in particular with the Wightman positivity condition for composite fields (see Section 5. D, below), one should demand that $n(\chi)$ has neither poles nor zeros in the right half plane and is positive for $c > 0$ ($h \geq 1$). If we assume in addition that $n(\chi)$ is polynomially bounded at infinity (in any direction different from the negative c-axis) we end up with the following (unique) choice (cf. [K5]):

$$n_C(\chi) = (h+\ell+c-1)\frac{\Gamma(h+c-1)}{\Gamma(c)}. \qquad (5.26)$$

A third choice is appropriate in the study of exceptional points for which h+c-2 is a negative integer. In that case we shall use

$$n_-(\chi) = n_C(\tilde{\chi}) = (h+\ell-c-1)\frac{\Gamma(h-c-1)}{\Gamma(-c)}. \qquad (5.26')$$

We shall use the notation G^{+}, $G^{\bullet}$, G and G^{-} for the intertwining operator with normalization (5.24), (5.25), (5.26) and (5.26'), respectively.

We remark that the product $n(\chi)\, n(\tilde{\chi})$ is the same for all four different choices of $n(\chi)$ (in fact, it remains the same for any $n(\chi)$ consistent with (4.11)). Knapp and Stein [K3] have shown that this product is proportional to the Plancherel measure for G (see Section 8.A., below).

We are now prepared to study the positivity properties of the p-space intertwining kernel

$$G^{+}_{\chi}(p) = (\tfrac{1}{2}p^{2})^{c} \sum_{s=0}^{\ell} \kappa_{\ell s}(c)\, \Pi^{\ell s}(p) = (\tfrac{1}{2}p^{2})^{c} \sum_{s=0}^{\ell} \frac{(h+s-c-1)_{\ell-s}}{(h+s+c-1)_{\ell-s}}\, \Pi^{\ell s}(p) \qquad (5.27)$$

(where $\Pi^{\ell s}$ are the projection operators (5.21) satisfying conditions (5.12)-(5.14)) and to single out the complementary series of unitary type I representations of $O^{\uparrow}(2h+1, 1)$. Since the projection operators are positive semidefinite (having eigenvalues 0 and 1), it is enough to control the sign of the coefficients $\kappa_{\ell s}$ and of the distribution $(p^{2})^{c}$ on the space $C_{\tilde{\chi}}$ (of the Fourier transforms of functions of $C_{\tilde{\chi}}$).

Let us start with the case $\ell = 0$. In this case,

$$\kappa_{00}(c) = 1 \left(= \Pi^{00}(p)\right) \qquad (5.23c)$$

and the right hand side of (5.27) looks superficially positive for all real c. However, the distribution $(p^2)^c$ requires regularization which destroys positivity for $c \leq -h$. On the other hand, the positivity of $G^+_\chi(p)$ guarantees the positivity of the scalar product

$$\int\int \overline{f(x_1)}\, G^+_\chi(x_1 - x_2)\, f(x_2)\, dx_1\, dx_2 \text{ for } f(x) \in C_\chi$$

only if the Fourier transform of f(x) exists and is square integrable with weight $G^+_\chi(p)$ and Eq. (5.7) makes sense. Now f(x) is infinitely differentiable so that f(p) falls off fast for $p \to \infty$. The only trouble may arise from the behavior of f(x) at infinity (or of f(p) at the origin). According to (2.29) the asymptotic behavior of a function $f(x) \in C_{\tilde{\chi}}$ (for $\chi = [0, c]$) is given by

$$f(x) \sim \frac{A}{(x^2)^{h-c}} \qquad \text{for } x \to \infty .$$

This implies that f(p) behaves at worst as $(p^2)^{-c}$ (for $c \geq 0$, $p^2 \to 0$). Then the positivity condition is ensured by the existence of the integral

$$\int_{p^2<\Lambda^2} G^+_\chi(p)\, |f(p)|^2 d^{2h}p = A \int_{p^2<\Lambda^2} (p^2)^c (p^2)^{-2c} d^{2h}p = A \int_0^{\Lambda^2} \frac{dp^2}{(p^2)^{c-h+1}} \quad (< \infty) \qquad (5.28)$$

which is satisfied for $c < h$. Thus, the representation $\chi = [0, c]$ is unitary for $-h < c < h$. (Note that this was the first place where the special choice of normalization (5.24) did matter; for the choice $n = n_C$ (5.26), G_χ would have poles for positive integers c's including the points $c = 1, 2, \ldots, h-1$ in the middle of the complementary series, and would change sign in each such point.)

Similarly one proves that for $\ell > 0$, $h > 1$, the sesquilinear form (5.7) on $C_{\tilde{\chi}} \times C_{\tilde{\chi}}$ defined by G^+_χ is positive defininite if $1-h < c < h-1$. The integer points (3.5), in particular the boundary points for the complementary series, require a special investigation which will be postponed to Chapter III.

The results of this section can be summarized by the following

Theorem 5.1. The sesquilinear form (5.7) on $C_{\tilde{\chi}} \times C_{\tilde{\chi}}$ with kernel G^+_χ given by (4.19)(5.24) (with Fourier transform (5.17)) is positive definite for

$$\ell = 0 \qquad -h < c < h \qquad (h \geq 1) \tag{5.29a}$$

$$\ell = 1, 2, \ldots \qquad 1-h < c < h-1 \qquad (h > 1). \tag{5.29b}$$

The unitary representations of $G_{(ex)}$ χ and $\tilde{\chi}$ in the domain (5.29) (which differ only in the sign of c) are equivalent.

The unitary IR χ in the range (5.29) with $c \neq 0$ are called the complementary series of type I unitary representations of $O^\uparrow(2h+1, 1)$.

5. D Wightman positivity

It turns out that the harmonic expansion (5.23) is also suited for the study of Wightman (Minkowski space) positivity of the 2-point function in a (weakly) conformal covariant quantum field theory.

The problem has also a purely mathematical formulation--and justification (without explicit reference to quantum field theory) and it would be in the spirit of the present work to emphasize that aspect of the matter.

It was shown recently (see [L2]) that for some range of real c the elementary representations of $SO^{\uparrow}(2h+1, 1)$ can be "continued" (in a peculiar way) to a special class of unitary representations of the universal covering G(2h, 2) of the conformal group $SO_0(2h, 2)$. The maximal compact subgroup $SO(2h) \otimes SO(2)$ of this group has a continuous center--the subgroup SO(2). The unitary representations of G(2h, 2), obtained via analytic continuation of the elementary representations of G, can be characterized by the requirement that the (hermitian) generator H of this distinguished SO(2) subgroup (which plays the role of a "conformal Hamiltonian," see [S5]) is non-negative. This is precisely the interpolated holomorphic discrete series of unitary IR of G (2h, 2), studied for instance in [R2, G10, 02]. The subsequent analysis of Wightman positivity could therefore be regarded as determination of the range of χ for which the corresponding representations of G(2h, 2) are unitary. It also appears to be equivalent to studying the Osterwalder-Schrader positivity (see [M3]).

Here we start with the "Euclidean 2-point Green function" (5.23) with normalization (5.26) and continue to Minkowski space momenta $p_{2h} = ip_0$ (p_0 - real). The Wightman 2-point function for composite fields in momentum space $w_C(p)$ is obtained from (5.23) (5.26) in two steps. First, we define the Minkowski space τ-function by replacing $(p^2)^c$ in (5.23) by $(p^2 - i0)^c = (\underline{p}^2 - p_0^2 - i0)^c$:

$$\tau_\chi(p) = \frac{\Gamma(-c)}{\Gamma(c)} (\tfrac{1}{2}p^2 - i0)^c \sum_{s=0}^{\ell} \kappa_{\ell s}(c)\, \Pi^{\ell s}(p). \tag{5.30}$$

(Here and in what follows we use the notation and generalized functions' techniques of Gel' fand et. al. [G4] .) Then we determine w(p) in terms of the discontinuity of τ_χ:

$$w_C(p) = -i\,\theta(p_0)\,[\tau_\chi(p) - \tilde{\tau}_\chi(p)] =$$

$$= -2\sin\pi(c-\ell)\frac{\Gamma(-c)}{\Gamma(c)}\,\theta(p_0)(-\tfrac{1}{2}p^2)_+^{c-\ell}\sum_{s=0}^{\ell}\kappa_{\ell s}(c)(\tfrac{1}{2}p^2)^\ell\,\Pi^{\ell s}(p) =$$

$$= \frac{2\pi\,\theta(p_0)}{\Gamma(c)\Gamma(c+1)}(-\tfrac{1}{2}p^2)_+^{c-\ell}\sum_{s=0}^{\ell}\frac{(c+2-h-\ell)_{\ell-s}}{(c+h+s-1)_{\ell-s}}(-1)^s(\tfrac{1}{2}p^2)^\ell\,\Pi^{\ell s}(p). \tag{5.31}$$

In deriving the last equality, we used the distribution theoretic identity

$$(Q + i0)^{\lambda} - (Q - i0)^{\lambda} = 2i \sin \pi\lambda \ (-Q)_{+}^{\lambda} \qquad (5.32)$$

where $t_{+}^{\lambda} = \theta(t)t^{\lambda}$ (see [G4]). It can be verified that the distribution $w_C(p)$ (5.31) is the Fourier transform of the x-space Wightman function, defined, as usual, as the boundary value

$w_C(x) = \lim_{\varepsilon \downarrow 0} G_{\chi} (\underline{x}, - \varepsilon - ix_0)$.

In order to find the range of positivity of the right-hand side of (5. 31), we first notice that the operator

$$(-1)^s (\tfrac{1}{2}p^2)^{\ell} \Pi^{\ell s} (p)$$

is Minkowski space positive, since, according to (5.17) (5.19), for $\mathfrak{z}_1 = \mathfrak{z} = \bar{\mathfrak{z}}_2$,

$$(-1)^s (\tfrac{1}{2}p^2)^{\ell} \Pi^{\ell s}(p) = A_{\ell s} |p \ |^{2\ell} C_s^{h-\frac{3}{2}} (\omega) \geq 0 \qquad (5.33)$$

for $\omega = 1 - p^2 \dfrac{\mathfrak{z}\bar{\mathfrak{z}}}{|p\mathfrak{z}|^2} \geq 1.$

The last inequality ($\omega \geq 1$) is fulfilled, because we have

$$\mathfrak{z}\bar{\mathfrak{z}} = \underline{\mathfrak{z}}\underline{\bar{\mathfrak{z}}} - \mathfrak{z}_0\bar{\mathfrak{z}}_0 > 0 \ \text{for} \ \mathfrak{z}^2 = 0 \ (= \bar{\mathfrak{z}}^2) \qquad (5.34)$$

and $p^2 = \underline{p}^2 - p_0^2 < 0$ (because of the $\theta(-p^2)$ factor). Now it is quite straightforward to establish the following positivity condition for the composite field Wightman functions

$$w_c(p) \geq 0 \text{ if } c \geq 0,\ \ell = 0, \text{ or } c \geq h + \ell - 2,\ \ell = 1, 2, \ldots. \tag{5.35}$$

The limit case $c = h + \ell - 2$ ($\ell = 1, 2, \ldots$) corresponds to the conserved "canonical currents." The composite character of the fields is reflected in the restriction $c \geq 0$. If we wish to incorporate elementary fields it is appropriate to use the normalization

$$n_w(\chi) = 2^c (h+\ell+c-1)\,\Gamma(h+c-1), \tag{5.36}$$

which violates condition (4.11). (The essential distinction from (5.26) is, of course, the absence of the factor $[\Gamma(c)]^{-1}$.) With this choice the Wightman function assumes the form:

$$w(p) = 2\pi \frac{\theta(p_0)}{\Gamma(c+1)} (-p^2)_+^{c-\ell} \sum_{s=0}^{\ell} \frac{(c+2-h-\ell)_{\ell-s}}{(c+h+s-1)_{\ell-s}} (-1)^s (p^2)^\ell \Pi^{\ell s}(p), \tag{5.37}$$

which is positive in the wider range $c \geq -1$ for $\ell = 0$. The limit case $\ell = 0$, $c = -1$ reproduces the two-point function for a free, spinless, zero mass field and (5.36) is adapted to the canonical normalization of such fields.

The positivity restriction on the "anomalous dimension" $h + c$ of (weakly) conformal covariant quantized fields was also obtained by other methods in refs. [R2], [F2].

III. PROPERTIES OF ELEMENTARY REPRESENTATIONS AT EXCEPTIONAL INTEGER POINTS

6. Nondecomposable representations and intertwining differential operators

6.A Subrepresentations of exceptional elementary representations

The present chapter will be devoted to the study of the four families of exceptional type I representations, $\chi^{\pm}_{\ell\nu} = [\ell, \pm(h+\ell+\nu-1)]$, $\chi'^{\pm}_{\ell\nu} = [\ell+\nu, \pm(h+\ell-1)]$. We shall use the shorthand notation $C^{(')\pm}_{\ell\nu}$ instead of $C_{\chi^{\pm}_{\ell\nu}}$ or $C_{\chi'^{\pm}_{\ell\nu}}$.

As was pointed out in Section 3 A the space $C^{-}_{\ell\nu}$ contains a finite dimensional invariant subspace $E_{\ell\nu}$, which can be described in the (x, ζ)-picture as follows. It consists of all polynomials $P(x,\zeta) \in C^{-}_{\ell\nu}$ $(x \in R^{2h}, \zeta^2 = 0)$ satisfying the ν-th order differential equation displayed below:

$$E_{\ell\nu} = \left\{P(x,\zeta) \in C^{-}_{\ell\nu},\ P\text{ - polynomial};\ (\zeta\nabla)^{\nu} P(x,\zeta) = 0 \text{ for } \nabla = \frac{\partial}{\partial x}\right\}. \tag{6.1}$$

This and similar statements in the rest of this subsection follow from the intertwining properties of the corresponding differential operators which are established in Section 6 B., below. The general solution of (6.1) can be written as a finite expansion of the type (2.29):

$$P(x,\zeta) = \sum_{k=0}^{\nu-1} (x^2)^{\nu-k-1}\, h_k(x; \{\zeta, x\}) \tag{6.2a}$$

where $\{\zeta, x\}$ stands for the 4-tuple

$$\{\zeta, x\} = (\zeta,\ x\zeta,\ x\wedge\zeta,\ 2(x\zeta)\,x - x^2\zeta). \tag{6.2b}$$

$$(x\wedge\zeta)_{\mu\nu} = x_\mu \zeta_\nu - \zeta_\mu x_\nu .$$

and h_k is a harmonic polynomial of degree k in the first argument and a homogeneous polynomial of degree ℓ in the last four variables:

$$\Delta_x h_k(x; \{\zeta, x'\}) = 0, \quad (\zeta\partial_\zeta - \ell) h_k\, (x; \{\zeta, x\}) = 0. \tag{6.2c}$$

Note that the four linear forms (6.2b) of ζ satisfy the first order differential equation

$$(\zeta\nabla)\ \{\zeta, x\} = 0. \tag{6.3}$$

They are in one-to-one correspondence with the generators of G displayed in Section 3.D if we identify ζ with the (infinite dimensional) generator $-\nabla$ of translation (see (3.29)), then $x\zeta$ would correspond to dilatations, $x\wedge\zeta$ would represent SO(2h) rotations and $2x(x\zeta) - x^2\zeta$ would correspond to special conformal transformations.

The space $C^+_{\ell\nu}$ contains an infinite dimensional invariant subspace

$$F_{\ell\nu} = \{ f(x,\zeta) \in C^+_{\ell\nu}\ (\Delta_\zeta f(x,\zeta) = 0);\ \exists g(x,\zeta) \in C'^+_{\ell\nu}:$$

$$f(x,\zeta) = (\partial\nabla)^\nu\ g(x,\zeta)\} \tag{6.4}$$

$$\partial = \frac{\partial}{\partial\zeta}, \quad \nabla = \frac{\partial}{\partial x} .$$

(The notation for the subspaces $E_{\ell\nu}$ and $F_{\ell\nu}$ is taken from ref. [G2], where the integer points in the space of SL(2, C) representations are studied.)

The subspaces $E_{\ell\nu}$ and $F_{\ell\nu}$ are orthogonal with respect to the invariant bilinear form (3.25) on $C^-_{\ell\nu} \times C^+_{\ell\nu}$. Indeed, if $P \in E_{\ell\nu}$ and $f \in F_{\ell\nu}$, then

$$B(P,f) = \frac{1}{\ell!} \int P(x, \partial_\zeta) f(x,\zeta) dx = \frac{1}{\ell!} \int P(x,\partial_\zeta)(\partial_\zeta \nabla)^\nu g(x,\zeta) dx =$$

$$= (-1)^\nu \frac{1}{\ell!} \int [(\partial_\zeta \nabla)^\nu P(x,\partial_\zeta)] \, g(x,\zeta) \, dx = 0. \tag{6.5}$$

(We note that this calculation is only legitimate if both P and g are harmonic in their second argument.)

The space $C'^-_{\ell\nu}$ contains an infinite dimensional invariant subspace

$$F'_{\ell\nu} = \{ f \in C'^-_{\ell\nu} ; \exists \, g \in C^-_{\ell\nu} : \; f(x,\zeta) = (\zeta\nabla)^\nu g(x,\zeta) \}. \tag{6.6}$$

Its dual, $C'^+_{\ell\nu}$ contains another infinite dimensional invariant subspace (for $h > 1$):

$$D_{\ell\nu} = \{ f(x,\zeta) \in C'^+_{\ell\nu} \; ; \; (\partial_\zeta \nabla)^\nu f(x,\zeta) = 0 \}. \tag{6.7}$$

The subspaces $D_{\ell\nu}$ and $F'_{\ell\nu}$ are orthogonal with respect to the invariant bilinear form (3.25). If $f(x, \zeta)$ is the harmonic extension of $(\zeta\nabla)^{\nu} g(x,\bar{\zeta}) \in F'_{\ell\nu}$, and $f' \in D_{\ell\nu}$, then

$$B(f, f') = \frac{1}{(\ell+\nu)!} \int f(x, \partial_{\zeta}) f'(x, \zeta)\, dx = 0 \qquad (6.5')$$

The proof is the same as in (6.5).

Conversely, if for fixed $f^{(\prime)}$ in $C^{(\prime)+}_{\ell\nu}$ Eq. $(6.5^{(\prime)})$ holds for any $P \in E_{\ell\nu}$ $(g \in C^{-}_{\ell\nu})$ then $f \in F_{\ell\nu}$ $(f' \in D_{\ell\nu})$. Thus, one can use (6.5) and (6.5′) as alternative definitions of $F_{\ell\nu}$ and $D_{\ell\nu}$. Similarly, assuming that $F_{\ell\nu}$ and $D_{\ell\nu}$ are defined by (6.4) and (6.7), we can use (6.5) and (6.5′) to define $E_{\ell\nu}$ and $F'_{\ell\nu}$.

Remark: In the special case 2h = 3, the exceptional representation $\chi'^{+}_{\ell\nu} = [\ell + \nu, \frac{1}{2} + \ell]$ of $SO^{\uparrow}(4,1)$ is still reducible when restricted to the subspace $D_{\ell\nu} \subset C'^{+}_{\ell\nu}$. This is most easily seen by exhibiting the K content of $D_{\ell\nu}$. According to Corollary 2.1 of the reciprocity theorem of Section 2.C an IR of SO(4) with highest weight $\tau = (\tau_1, \tau_2)$ is contained in $\chi'^{+}_{\ell\nu}$ iff τ_1 and τ_2 are integers and $|\tau_1| \leq \ell + \nu \leq \tau_2$. On the other hand, according to (6.4) (6.7) the factor space $C'^{+}_{\ell\nu}/D_{\ell\nu}$ is isomorphic to $F_{\ell\nu}$ $(\subset C^{+}_{\ell\nu})$ and hence has the same K-content which is given by the representations $\tau = (\tau_1, \tau_2)$ such that $|\tau_1| \leq \ell$, $\tau_2 \geq \ell + \nu$. Thus the K-content of $D_{\ell\nu}$ splits in two disjoint sets:

$$D_{\ell\nu}[SO^{\uparrow}(4,1)] = D^{+}_{\ell\nu} \times D^{-}_{\ell\nu},$$

where

$$D^+_{\ell\nu} = \bigoplus_{i=1}^{\nu} \bigoplus_{k=0}^{\infty} (\ell+i,\ \ell+\nu+k)\ ,$$

$$D^-_{\ell\nu} = \bigoplus_{i=1}^{\nu} \bigoplus_{k=0}^{\infty} (-\ell-i,\ \ell+\nu+k) \qquad (6.8)$$

There are no elements in the Lie algebra of $SO^{\uparrow}(4,1)$ which would connect $D^+_{\ell\nu}$ and $D^-_{\ell\nu}$. However, according to the results quoted in Sec. 2A $D^{\pm}_{\ell\nu}$ are mirror images of one another. Therefore, the direct sum space $D_{\ell\nu}$ is irreducible with respect to the extended group $O^{\uparrow}(4,1)$ (which includes space reflections).

We note that in the case of the Lorentz group (i.e. for $h=1$) $F'_{\ell\nu} = C'^-_{\ell\nu}$ and $D_{\ell\nu} = 0$ for $\ell \geq 1$ (see Appendix B). For $h>1$ all the subspaces $I^{(')\pm}_{\ell\nu}$ are non-trivial for any ℓ.

6.B Intertwining differential operators. Partial equivalence among the representations $\chi^{(')\pm}_{\ell\nu}$

Proposition 6.1. The differential operators

$$d^{\nu} = (\zeta\nabla)^{\nu} :\quad C^-_{\ell\nu} \to C'^-_{\ell\nu} \qquad (6.9a)$$

and

$$d'^{\,\nu} = (\partial_{\zeta}\nabla)^{\nu} :\quad C'^+_{\ell\nu} \to C^+_{\ell\nu} \qquad (6.9b)$$

are intertwining operators for the pairs of representations $(\chi^-_{\ell\nu}, \chi'^-_{\ell\nu})$ and $(\chi'^+_{\ell\nu}, \chi^+_{\ell\nu})$, respectively:

$$d^{\nu} T_{\chi^-_{\ell\nu}} = T_{\chi'^-_{\ell\nu}} d^{\nu} \tag{6.10a}$$

$$d'^{\nu} T_{\chi'^+_{\ell\nu}} = T_{\chi^+_{\ell\nu}} d'^{\nu} . \tag{6.10b}$$

(The operator d'^{ν} is always assumed to act on harmonic functions of ζ, so that we can set $\partial^2_{\zeta} = 0$.)

The simplest way to prove this proposition is to show that the operators d^{ν} and d'^{ν} have the right commutation relations with the generators of the elementary representations exhibited in Sec. 3D. Verification of the infinitesimal form of (6.10) is only nontrivial for dilatations and special conformal transformations. For the dilatations (3.33) it is sufficient to observe that

$$(q\nabla)^{\nu}(\ell+\nu-1-x\nabla) = (\ell-1-x\nabla)(q\nabla)^{\nu}, \quad q = \mathfrak{z} \text{ or } \partial_{\zeta} . \tag{6.11}$$

To verify (6.10) for the special conformal transformations (3.34) (with c taken from (3.5)) one uses identities of the type

$$[(q\nabla)^{\nu}, x_{\mu}] = \nu q_{\mu}(q\nabla)^{\nu-1}, \quad [(q\nabla)^{\nu}, x^2] = 2\nu(xq)(q\nabla)^{\nu-1},$$

$$[(q\nabla)^{\nu}, x_{\mu}(x\nabla)] = \nu[x_{\mu}(q\nabla) + (\nu-1)q_{\mu} + q_{\mu}(x\nabla)](q\nabla)^{\nu-1}.$$

In deriving the second equality we have used that $q^2 = 0$ for both $q = \mathfrak{z}$ and $q = \partial_{\zeta}$.

For exceptional χ, the dual representations χ and $\tilde{\chi}$ are no longer equivalent. In that case the intertwining operators

We have, in particular, the identities

$$d^{\nu} G^{-}_{\ell\nu} = (\mathfrak{z}\nabla)^{\nu} G^{-}_{\ell\nu}(x;\mathfrak{z},\zeta) = 0 = G^{+}_{\ell\nu} G^{-}_{\ell\nu}, \tag{6.15a}$$

$$d'^{\nu} G'^{+}_{\ell\nu} = (\partial_{\zeta}\nabla)^{\nu} G'^{+}_{\ell\nu}(x;\zeta,\mathfrak{z}) = 0 = G'^{-}_{\ell\nu} G'^{+}_{\ell\nu}, \tag{6.15b}$$

$$G'^{+}_{\ell\nu} d^{\nu} = G'^{+}_{\ell\nu}(p;\zeta,\partial_{\mathfrak{z}})(i\mathfrak{z}p)^{\nu} = 0 = G'^{+}_{\ell\nu} G'^{-}_{\ell\nu}, \tag{6.15c}$$

$$G^{-}_{\ell\nu}\, d'^{\nu} = G^{-}_{\ell\nu}(p;\mathfrak{z},\partial_{\zeta})(i\partial_{\zeta}p)^{\nu} = 0 = G^{-}_{\ell\nu} G^{+}_{\ell\nu}. \tag{6.15d}$$

<u>Proof.</u> The intertwining kernel

$$G^{-}_{\ell\nu}(x;\mathfrak{z},\zeta) = \frac{n_{-}(\chi^{-}_{\ell\nu})}{(2\pi)^{h}}\left(\frac{x^2}{2}\right)^{\nu+\ell-1}(-\mathfrak{z}\, r(x)\zeta)^{\ell} \tag{6.16}$$

is a polynomial in x of the type (6.2) and is, therefore, annihilated by d^{ν}. On the other hand $G^{+}_{\ell\nu}(p)$ is a homogeneous function of p of degree $2(\ell+\nu+h-1)$, which annihilates Fourier transforms of polynomials $P \in E_{\ell\nu}$ [since $\mathcal{F}\, P(p,\mathfrak{z}) = P(i\nabla_{p},\mathfrak{z})\delta(p)$ and the degree of $P(x,\mathfrak{z})$ does not exceed $2(\nu+\ell-1)$, cf. (6.14a]. This takes care of (6.15a).

To prove the first equation (6.15b) we use (6.13) and (5.14) which can be written as

$$(p\partial_{\zeta})^{\nu}\, \Pi^{\ell+\nu}\, \overset{s}{\ }(p;\zeta,\mathfrak{z}) = 0 \qquad \text{for } s = \ell+1,\ldots,\ell+\nu \tag{6.17}$$

The last equation (6.15b) is a consequence of the orthogonality property of projection operators, since

$$G'^{-}_{\ell\nu}(p) = \frac{\ell!}{\nu!}(2h+\ell-3)!\,(2h+2\ell+\nu-2)!\,\left(\frac{2}{p^2}\right)^{h+\ell-1}\sum_{s=0}^{\ell}\frac{(-1)^s\,\Pi^{\ell+\nu s}(p)}{(\ell-s)!\,(2h+\ell+s-3)!} \qquad (6.18)$$

Eqs. (6.15c) and (6.15d) now follow because of the symmetry of the kernels $G_{\ell\nu}(.;\zeta_1,\zeta_2)$ with respect to the last two arguments.

Proposition 6.2 implies a number of isomorphisms (and equivalences of the corresponding IR's of G_{ex}) among the factor spaces C_χ/I_χ for each exceptional (integer) χ and the invariant subspaces of the partially equivalent representations:

$$F_{\ell\nu} \simeq C^-_{\ell\nu}/E_{\ell\nu} \simeq F'_{\ell\nu} \simeq C'^+_{\ell\nu}/D_{\ell\nu}\,,\quad E_{\ell\nu} \simeq C^+_{\ell\nu}/F_{\ell\nu}\,,\quad D_{\ell\nu} \simeq C'^-_{\ell\nu}/F'_{\ell\nu}\,. \qquad (6.19)$$

We remark that the exact sequences of the quartet diagram of Fig. 1 are parts of longer exact sequences relating by partial equivalences all exceptional elementary representations (3.5) with the same infinitesimal character [Z5, G1] (and not just the type I representations, considered here). We also note that similar diagrams of intertwining mappings are known for the complex semi-simple Lie groups (see [G2] [Z3]).

6.C Hermitian forms on invariant subspaces. Exceptional series of unitary representations

As was observed in the preceding subsection the operators G^+_χ (with normalization factor n_+ (5.24)) are not defined on the spaces

$C^{(\prime)-}_{\ell\nu}$. Similarly, the operators G^-_χ (with normalization n_- (6.14)) become infinite on the spaces $C^{(\prime)+}_{\ell\nu}$. We shall prove here that these singular operators can nevertheless be defined as quadratic forms on invariant subspaces.

We start with a heuristic discussion of the problem, referring, for the sake of definiteness, to G^+. The kernel $G^+_\chi(x)$ differs from the kernel $G^-_\chi(x)$ (which is well defined for all c in the left half plane) by the numerical factor

$$\frac{n_+(\chi)}{n_-(\chi)} = \frac{(h+\ell+c-1)\,\Gamma(h+c-1)}{(h+\ell-c-1)\,\Gamma(h-c-1)} . \tag{6.20}$$

Let us use the uniform notation $I^{(\prime)\pm}_{\ell\nu}$ for the invariant subspaces of $C^{(\prime)\pm}_{\ell\nu}$. The operator G^-_χ vanishes on the invariant subspaces $I^{(\prime)-}_{\ell\nu}$ of $C^{(\prime)-}_{\ell\nu}$ (for $\chi = \chi^{(\prime)-}_{\ell\nu}$), while the numerical factor (6.20) becomes infinite for exceptional representations with negative c. We have to show that their product G^+_χ defines a non-vanishing (and finite) bilinear form on $I^{(\prime)-}_{\ell\nu} \times I^{(\prime)-}_{\ell\nu}$. The most natural way to do that is to use a limiting procedure (reminiscent to analytic renormalization).

Let, for example, $\chi^-_\varepsilon = \chi^-_{\ell\nu}(\varepsilon) = [\,\ell, 1-k-\ell-\nu+\varepsilon]$;

consider the hermitian form

$$B^+_{\ell,\nu,\varepsilon}(f_1, f_2) = \iint \langle f_1(x_1), G^+_{\chi^-_\varepsilon}(x_1-x_2) f_2(x_2)\rangle \, dx_1 dx_2 , \quad f_1, f_2 \in F_{\ell\nu} , \tag{6.21}$$

which is well defined for positive ε. In order to find the limit of (6.21) for $\varepsilon \to 0$ we expand the kernel $G^+_{\chi^-_\varepsilon}$ in powers of ε:

$$G^+_{\chi^-_\varepsilon}(x;\, ,\zeta) = \frac{(-1)^\nu}{(2\pi)^h} \frac{\nu\,(x^2/2)^{\nu+\ell-1}}{(\ell+\nu)!\,\Gamma(h+\ell+\nu-1)} \left[\frac{1}{\varepsilon} - \log\frac{x^2}{2} + O(\varepsilon)\right] (\mathfrak{z}\, r(x)\zeta)^\ell \qquad (6.22)$$

The crucial point is that the singular $(\frac{1}{\varepsilon})$ term in (6.21) vanishes because of (6.5) (since $(x^2)^{\nu+\ell-1}(\mathfrak{z}r(x)\zeta)^\ell \in E_{\ell,\nu}$ and $f_{1,2} \in F_{\ell\nu}$). Thus we can go to the limit $\varepsilon \downarrow 0$, obtaining*)

$$(f_1, B^+_{\ell\nu} f_2) \equiv \lim_{\varepsilon\downarrow 0} B^+_{\ell,\nu,\varepsilon}(f_1, f_2) = \iint f_1(x_1), B^+_{\ell\nu}(x_1 - x_2) f_2(x_2)\, dx_1\, dx_2, \qquad (6.23)$$

where

$$B^+_{\ell\nu}(x;\mathfrak{z},\zeta) = \frac{(-1)^\nu}{(2\pi)^h} \frac{\nu}{(\ell+\nu)!\,\Gamma(h+\ell+\nu-1)} \left(\frac{x^2}{2}\right)^{\nu-1} \log\frac{x^2}{2} \left[(x\mathfrak{z})(x\zeta) - \frac{x^2}{2}(\mathfrak{z}\zeta)\right]^\ell. \qquad (6.24)$$

The kernel $B^+_{\ell\nu}(x_{12};\mathfrak{z},\zeta)$ is not covariant under the representation $T^{\chi^+_{\ell\nu}}$ of $G_{(ex)}$. Under dilatations and special conformal transformations it acquires non-homogeneous terms (because of the log). For instance, under the conformal inversion (1.29) (2.30) (2.31c) $B^+_{\ell\nu}$ transforms as

$$B^+_{\ell\nu}(x_{12};\mathfrak{z}_1,\mathfrak{z}_2) \to (x_1^2\, x_2^2)^{\ell+\nu-1} B^+_{\ell\nu}(Rx_1, -Rx_2; r(x_1)\mathfrak{z}_1, r(x_2)\mathfrak{z}_2) =$$

$$= B^+_{\ell\nu}(x_{12};\mathfrak{z}_1,\mathfrak{z}_2) +$$

$$+ \frac{(-1)^{\nu+1}}{(2\pi)^h} \frac{\nu(\log x_1^2 + \log x_2^2)}{(\ell+\nu)!\,\Gamma(h+\ell+\nu-1)} \left(\frac{x_{12}^2}{2}\right)^{\nu-1} \left[(x_{12}\mathfrak{z}_1)(x_{12}\mathfrak{z}_2) - \tfrac{1}{2}x_{12}^2(\mathfrak{z}_1\mathfrak{z}_2)\right]^\ell. \qquad (6.25)$$

---

*) We note that $(f_1, B^+_{\ell\nu} f_2) = \left[\frac{d}{d\varepsilon}\left(\varepsilon B^+_{\ell,\nu,\varepsilon}(f_1,f_2)\right)\right]_{\varepsilon=0}$ (6.23')
(see (6.21)), because

$$\frac{d}{d\varepsilon}\left(\varepsilon G^+_{\chi^-_\varepsilon}\right)\Big|_{\varepsilon=0} = \frac{1}{(2\pi)^h}(-1)^{\nu+1}\nu(\mathfrak{z}\, r(x)\zeta)^\ell \left(\frac{x^2}{2}\right)^{\ell+\nu-1}\left[\log\frac{x^2}{2} - \psi(\ell+\nu+1) - \psi(h+\ell+\nu+1) + \frac{1}{\nu}\right] \qquad (6.24')$$

where $\psi(z) = \frac{d}{dz}\ln\Gamma(z)$. The last three terms in the square brackets do not contribute to the right-hand side of (6.23') for the same reason as the term above.

However, it is not difficult to see that the hermitian form (6.23) remains invariant, since the inhomogeneous terms can be split (as exhibited by (6.25)) into a sum of polynomials of $E_{\ell\nu}$ in one of the variables (multiplied by more complicated functions of the other variable) and therefore have vanishing expectation values (6.23) for $f_1, f_2 \in F_{\ell\nu}$.

Let I_χ be the invariant subspace of C_χ for any of the (type I) exceptional representations (3.3). In general, we define the hermitian form B_{I_χ} on $I_\chi \times I_\chi$ by

$$(f_1, B_{I_\chi} f_2) = \lim_{\varepsilon \downarrow 0} \iint \langle f_1(x_1), G^{\pm}_{\tilde{\chi}_\varepsilon}(x_{12}) f_2(x_2) \rangle dx_1 dx_2 , \qquad f_1, f_2 \in I_\chi , \tag{6.26}$$

where $\chi_\varepsilon^{(\prime)\pm} = [\ell^{(\prime)}, \pm(c^{(\prime)} - \varepsilon)]$ for $\chi^{(\prime)\pm} = [\ell^{(\prime)}, \pm c^{(\prime)}]$ $(c^{(\prime)} > 0)$.

One finds by inspection that they are all finite (non-vanishing) and invariant under the corresponding representation of $O^{\uparrow}(2h+1,1)$. The explicit forms of their kernels look simpler in the x-space picture for c>0 and in the p-space picture for c<0; in particular, we have,

$$B'^{+}_{\ell\nu}(x;\mathfrak{z},\zeta) = \frac{(-1)^{\nu+1}}{(2\pi)^h} \frac{\nu \log(\frac{x^2}{2})}{\ell!\Gamma(h+\ell-1)} (\frac{x^2}{2})^{\ell-1} [\mathfrak{z} r(x)\zeta]^{\ell+\nu} , \tag{6.27}$$

$$B'^{-}_{\ell\nu}(p) = \frac{\nu!(p^2/2)^{h+\ell-1}}{\ell!(2h+\ell-3)!(2h+2\ell+\nu-2)!} \{\sum_{s=0}^{\ell}(-1)^s (2h+\ell+s-3)!(1-s)!\, \Pi^{\ell+\nu}_{s}(p) +$$

$$+(\log\frac{2}{p^2}) \sum_{s=\ell+1} (-1)^{\ell} \frac{(2h+\ell+s-3)!}{(s-\ell-1)!} \Pi^{\ell+\nu}_{s}(p)\} . \tag{6.28}$$

The definition of $B_{\ell\nu}^{(')\pm}$ implies the following

Proposition 6.3. *The hermitian form $B_{\ell\nu}^{(')\pm}$ on $I_{\ell\nu}^{(')\pm}$ is related to the hermitian form $(F_1, G_{\ell\nu}^{(')\pm} F_2)$ on $C_{\ell\nu}^{(')\pm}$ by*

$$(f_1, B_{\ell\nu}^{(')\pm} f_2) = (G_{\ell\nu}^{(')\pm} F_1, B_{\ell\nu}^{(')\pm} G_{\ell\nu}^{(')\pm} F_2) = (F_1, G_{\ell\nu}^{(')\pm} F_2) \qquad (6.29)$$

where $f_i = G_{\ell\nu}^{(')\pm} F_i$, $i = 1, 2$; $F_{1,2}$ can be regarded as cosets in the factor space

$$C_{\ell\nu}^{(')\pm} / I_{\ell\nu}^{(')\pm} (\ni F) \qquad (6.30)$$

(or as arbitrary representatives of these cosets). The second equation (6.29) can be written in the symbolic form

$$G_{\ell\nu}^{(')\pm} B_{\ell\nu}^{(')\pm} G_{\ell\nu}^{(')\pm} = G_{\ell\nu}^{(')\pm}, \qquad (6.29')$$

where both sides should be regarded as kernels of a hermitian form on $C_{\ell\nu}^{(')\pm}$ (or on the product of the factor spaces (6.30)).

According to this proposition and to the isomorphism (6.19) between the invariant subspaces I_χ and the dual factor spaces, the study of the positivity of the (singular) hermitian forms (6.26) is equivalent to the simpler problem of studying the positivity properties of the (regular) intertwining operators $G_{\ell\nu}^{(')\pm}$, defined in Sec. 6b.

There are three classes of subrepresentations to be studied: those realized in the subspaces $E_{\ell\nu}$, $F'_{\ell\nu} \simeq F_{\ell\nu}$ and $D_{\ell\nu}$.

The only finite dimensional unitary representation of $O^{\uparrow}(2h+1,1)$ is the trivial one (realized in the one dimensional space E_{01}). It corresponds to the end point ($\ell=0$, c = -h) of the complementary series of unitary representations (5.29a).

In order to single out the infinite dimensional unitary representations in the spaces $F_{\ell\nu}$ and $D_{\ell\nu}$ it is necessary to study the positivity properties of the p-space kernels (6.12) and (6.13). We see that $G^{+}_{\ell\nu}$ is only positive for $\ell = 0$, while $G'^{+}_{\ell\nu}(p)$ is positve for all ℓ and ν. The unitarity of the representations in $F_{0\nu}$ follows immediately; the proof of positivity of the scalar product in $D_{\ell\nu}$ requires an additional argument, which is presented in Appendix C. Thus we have

Proposition 6.4. The only unitary irreducible subrepresentations of the exceptional type I elementary representations $\chi^{(')\pm}_{\ell\nu}$ of $O^{\uparrow}(2h+1,1)$ are those realized in E_{01}, $F_{0\nu} \simeq F'_{0\nu}$ and $D_{\ell\nu}$.

We remark that the elementary representations $\chi'^{-}_{0\nu}$, containing the IR's $F'_{0\nu}$ are the limit points $\ell = \nu$, c = h-1 of the complementary series (5.29b). For 2h = 3 the representations in $D_{\ell\nu}$ form the discrete series of sqare integrable unitary representations of $O^{\uparrow}(4,1)$ which will be studied in Sec. 7 below. For integer $h \geq 2$ the representation in $D_{\ell\nu}$ is known [Z 5] to be equivalent to the irreducible principal series representation $[(1,1,\ldots,1,\ell+1,\ell+\nu),0]$ (which reduces to one of the representations (3.5c) for h = 2).

The unitary representations in $F_{0\nu}$ $(\simeq F'_{0\nu})$ and in $D_{\ell\nu}$ for all $2h \geq 5$ are called exceptional series of unitary representations of $O^{\uparrow}(2h+1,1)$.

A last remark concerns Thieleker assertion [T3] about the existence of yet another set of exceptional unitary (typeI) representations. These are the representations listed in case II.B.1) of Theorem 3 of [T3] with $\Lambda_{\mu_{j-1}} = n_j$, $j \geq 2$ (in Thieleker's notation). In the special case of type I IR's (j = 2) they would be contained in the elementary representations $\chi_\ell = [\ell, h+\ell-1]$, $\ell > 0$ (in the notation of the present paper). It appears that this assertion is not correct. Indeed, it follows from the results of Hirai [H3] and Knapp and Stein [K3, K4] that the representations χ_ℓ are topologically irreducible and that the only invariant hermitian form in C_{χ_ℓ} is the one generated by the intertwining operator $G_{\tilde{\chi}_\ell}$ considered here. On the other hand, Eq. (5.23) above shows that the kernel $G_{\tilde{\chi}_\ell}(p)$ is not positive definite. Therefore, we maintain that the representations χ_ℓ are (irreducible and) non-unitary.

6D Differential identities between hermitian forms for exceptional representations

The relation (6.29) between $B_{\ell\nu}$ and $G_{\ell\nu}$ is not the only one between the hermitian forms on invariant subspaces and equivalent factor spaces of exceptional elementary representations. The objective of this subsection is to establish two other sets of identities of that type, which involve the differential operators d^{ν} and d'^{ν}.

In order to write these new identities in a simple and symmetric form, we first notice that the singular bilinear forms $B^{+}_{\ell\nu}$ and $B'^{-}_{\ell\nu}$ are proportional to the appropriate limits of the intertwining operator G^{0}_{χ} normalized according to (5.25) for $\chi \to \chi^{-}_{\ell\nu}$ and $\chi \to \chi'^{+}_{\ell\nu}$, respectively. Indeed this follows from (6.26) and from the proportionality of the intertwining operators $G^{\pm}_{\chi}$ and G^{0}_{χ} (for non-exceptional χ). It appears more convenient for our present purposes to use G^{0}_{χ} rather than $B_{\tilde{\chi}}$.

Proposition 6.5. In the limit points $\chi^{-}_{\ell\nu}$ and $\chi'^{+}_{\ell\nu}$ (in which G^{0}_{χ} becomes infinite) the products $(p\mathfrak{z}_1)^{\nu} G^{0}_{\chi^{-}_{\ell\nu}}(p;\mathfrak{z}_1,\mathfrak{z}_2)(p\mathfrak{z}_2)^{\nu}$ and $(p\partial_{\zeta})^{\nu} G^{0}_{\chi'^{+}_{\ell\nu}}(p;\zeta,\partial_{\mathfrak{z}})(p\mathfrak{z})^{\nu}$ remain finite, and the following identities take place

$$(p\mathfrak{z}_1)^{\nu} G^{0}_{\ell\nu-}(p;\mathfrak{z}_1,\mathfrak{z}_2)(p\mathfrak{z}_2)^{\nu} = \frac{(2h+\ell-2)\nu}{(h+\ell-1)\nu} G'^{0}_{\ell\nu-}(p;\mathfrak{z}_1,\mathfrak{z}_2) \tag{6.31}$$

$$[\frac{\ell!}{(\ell+\nu)!}]^2 (p\partial_\zeta)^\nu G'^{o}_{\ell\nu+}(p;\zeta,\partial_z)(pz)^\nu = \frac{(2h+\ell-2)_\nu}{(h+\ell-1)_\nu} G^o_{\ell\nu+}(p;\zeta,\partial_z) \qquad (6.32)$$

where $G^{(\prime)o}_{\ell\,\nu\pm} = G^o_{\chi^{(\prime)\pm}_{\ell\nu}}$.

Proof. Eqs. (6.31) and (6.32) follow simply from similar identities for the projection operators:

$$(pz_1)^\nu \Pi^{\ell s}(p;z_1,z_2)(pz_2)^\nu = \frac{(\ell-s+1)_\nu (2h+\ell+s-2)_\nu}{(\ell+1)_\nu (h+\ell-1)_\nu} (\frac{p^2}{2})^\nu \Pi^{\ell+\nu\, s}(p;z_1,z_2), \qquad (6.33)$$

$$[\frac{\ell!}{(\ell+\nu)!}]^2 (p\partial_\zeta)^\nu \Pi^{\ell+\nu\, s}(p;\zeta,\partial_z)(pz)^\nu = \frac{(\ell-s+1)_\nu (2h+\ell+s-2)_\nu}{(\ell+1)_\nu (h+\ell-1)_\nu} (\frac{p^2}{2})^\nu \Pi^{\ell s}(p;\zeta,\partial_z) \qquad (6.34)$$

for $s = 0,\ldots,\ell$ $(\nu = 1,2,\ldots)$. The first of this equalities is a direct consequence of Eqs. (5.17) (5.19). The derivation of (6.34), on the other hand, uses the explicit form (5.21) of the harmonic extension of $\Pi^{\ell s}$, and requires some work.

First, we shall establish the following auxiliary relation for the functions $L_{\ell s}(p,\zeta)$, defined in (5.21b):

$$(\hat{p}\partial_\zeta) b_{\ell s} L_{\ell s}(p,\zeta) = \ell b_{\ell-1\, s} L_{\ell-1\, s}(p,\zeta) \text{ for } \ell \geq s+1, \quad \hat{p} = \frac{1}{\sqrt{p^2}} p. \qquad (6.35)$$

To prove (6.34) we notice that according to (5.21)

$$b_{\ell s} L_{\ell s}(p,\zeta) = \frac{\ell!}{s!}[(h+s-\tfrac{1}{2})_{\ell-s}]^{-1} (\zeta^2)^{\frac{\ell-s}{2}} P^{(h+s-\frac{3}{2},\, h+s-\frac{3}{2})}_{\ell-s}(\hat{p}\hat{\zeta}),$$

and we use a known identity (Eq. 8.961.3 of ref. [40]) for

Jacobi polynomials:

$$(\hat{p}\partial_\zeta)(\zeta^2)^{\frac{1}{2}n} P_n^{(\lambda,\lambda)}(\hat{p}\hat{\zeta}) = (\zeta^2)^{\frac{1}{2}(n-1)} \{n(\hat{p}\hat{\zeta})P_n^{(\lambda,\lambda)}(\hat{p}\hat{\zeta}) + [1-(\hat{p}\hat{\zeta})^2]\frac{d}{d(\hat{p}\hat{\zeta})} P_n^{(\lambda,\lambda)}(\hat{p}\hat{\zeta})\}$$

$$= (n+\lambda)(\zeta^2)^{\frac{1}{2}(n-1)} P_{n-1}^{(\lambda,\lambda)}(\hat{p}\hat{\zeta}).$$

Applying (6.35) ν times, we find

$$(\hat{p}\partial_\zeta)^\nu b_{\ell+\nu\, s} L_{\ell+\nu\, s}(p,\zeta) = \frac{(\ell+\nu)!}{\ell!} b_{\ell s} L_{\ell s}(p,\zeta) \quad (\text{for } s \leq \ell)$$

$$L_{\ell+\nu s}(p,\partial_z)(z\hat{p})^\nu = \frac{(\ell-s+1)_\nu (2h+\ell+s-2)_\nu}{2^\nu (h+\ell-1)_\nu} L_{\ell s}(p,\partial_z) \tag{6.36}$$

Inserting (6.36) in (5.21a) we obtain (6.34). This implies (6.32) since Eq. (6.17) guarantees the vanishing of the singular contribution (proportional to $\log \frac{2}{p^2}$) in the counterpart of (6.28).

7. Discrete series of unitary representations

7.A Definition and general properties of the discrete series of $SO^\uparrow(2n,1)$

Let $g \to U(g)$ be a unitary IR of a semisimple Lie group acting in a Hilbert space $\mathcal{H}$. It is said to belong to the discrete series if there is a non zero vector $\Psi \in \mathcal{H}$ for which

$$\int |(\Psi, U(g)\Psi)|^2 dg < \infty \tag{7.1}$$

where dg is (suitably normalized) invariant (Haar) measure on G. It is known ([W3] vol.I 4.5.9) that (7.1) implies the existence of a positive number d_U, called the formal dimension of U, such that

$$\int |(\varphi, U(g)\Psi)|^2 dg = d_U^{-1} \|\varphi\|^2 \|\Psi\|^2 ; \qquad (7.2)$$

(the value of d_U depends on the normalization of the Haar measure dg).

The question of existence of a discrete series is answered by the following theorem of Harish-Chandra [H2](Part II, Theorem 13).

Theorem 7.1. A semisimple Lie group G with finite center has a discrete series (of unitary IR) if and only if it admits a compact Cartan subalgebra (i.e. iff rank G = rank K).

For a proof the reader may also consult [W3] Vol. II, Theorem 10.2.1.2. (K is, as usual, the maximal compact subgroup of G).

Corollary 7.2. The groups $SO^{\uparrow}(N,1)$ and Spin(N,1) ($N \geq 2$) have a discrete series iff N is even (N = 2n).

This Corollary also follows from Theorem 9 of [M11] and from the observation that the restricted Weyl group of SO(N,1) contains the matrix $-\eta$ (see (1.1)).

A convenient characteristic of the discrete series representations is given in terms of the spherical trace function

$t_\tau^U(g) = \mathrm{Tr}(\Pi_\tau U(g) \Pi_\tau)$ (τ: IR of K = SO(2n) in $\mathcal{V}_\tau$; Π_τ: projection on $\mathcal{V}_\tau$)

introduced in Sec. 3.C. Let Θ_U be the character of a discrete series representation . Let further U be a subquotient of the elementary representation χ. Then according to Corollary 3.5 of Sec. 3.C. Θ_U is expressed in terms of the spherical trace function t_τ^χ:

$$\Theta_U(g) = \sum_{\tau \in U} t_\tau^\chi(g); \qquad t_\tau^\chi(g) = t_\tau^U(g) \text{ for } \tau \in U. \tag{7.3}$$

It can be shown that $t_\tau^\chi(g)$ is a spherical trace function of some discrete series representation iff it is square integrable on G.

There is a conjecture (Blattner's formula) which purports to give the K-content of the discrete series representations for an arbitrary semi-simple Lie group. This conjecture was recently proven to be true for the groups SO(2n, 1) by one of the authors [M4]. Here we shall restrict ourselves to the discrete series representations contained as subrepresentations (or quotients) of type I elementary representations of G. In this case the situation is simpler and our analysis can be based on the following

Theorem 7.3. The unitary irreducible discrete series representations U of $SO^\uparrow(2n, 1)$ and Spin (2n, 1) are not unitarily equivalent to their mirror image $\tilde{U}$ (defined in Sec. 2.A) Both U and $\tilde{U}$ appear as subrepresentations (more generally subquotients) of the same elementary representation.

The proof is based on the Harish-Chandra classification of discrete series representations for arbitrary semisimple Lie groups (see [H2, W3] and [M4]) and will be omitted.

Corollary 7.4. Discrete series representations of

<u>$G \simeq SO^{\uparrow}(2n,1)$ (or Spin(2n,1)) never contain a completely symmetric tensor representation of K (for $n \geq 2$).</u>

<u>Corollary 7.5.</u> <u>For $n \geq 3$ no discrete series representations of $G (\simeq SO^{\uparrow}(2n,1))$ occur as a subquotient of an elementary type I representation $\chi = [\ell, c]$ of G.</u>

To prove 7.4, we note that completely symmetric tensor IR's of K = SO(2n) are equivalent to their mirror image for $n \geq 2$. Therefore if such an IR τ of K occurs in a discrete series representation U, it will also occur in its (inequivalent) mirror image $\tilde{U}$. Theorem 7.3 then implies that τ will have to be contained twice in the elementary representation χ which contains both U and $\tilde{U}$ as subquotients. But that contradicts Corollary 2.1. of the (Frobenius) reciprocity theorem (see Sec. 2C).

To obtain 7.5 we recall that a representation $\tau = (\tau_1, \ldots, \tau_n)$ of K = SO(2n) $n \geq 3$ contains the type I representation $(\ell) = (0, \ldots 0, \ell_{n-1} = \ell)$ of M = SO(2n-1) only if $\tau_1 = \ldots = \tau_{n-2} = 0$ (see Sec. 2A). Such representations of K are equivalent to their mirror image and, therefore, we can apply to them the argument, which led us to 7.4.

Since the representation theory of $SO^{\uparrow}(2,1)$ is well known (see [B1] [G2] and Appendix B.4, below) Corollary 7.5 only leaves us with the group $SO^{\uparrow}(4,1)$ (and its covering Spin(4,1) which, however, contains no additional type I representations). In this case every (unitary) IR ℓ of $M \simeq SO(3)$ is of type I. According to the analysis of Sec. 6A Theorem 7.3 implies that the only candidates for the unitary discrete series IR's are the subrepresentations in the subspaces $D^{\pm}_{\ell\nu}$ (6.8) and their duals in $C'^{-}_{\ell\nu}$. In the following subsection we shall demonstrate that they indeed belong to the discrete series of $SO^{\uparrow}(4,1)$.

We shall close this general review by quoting a result of G. J. Zuckerman[Z4, Z5](specialized to our case).

Let $U_{\ell\nu}$ be the discrete series representation of $O^{\uparrow}(4,1)$ (acting in $D_{\ell\nu} \subset C'^{+}_{\ell\nu}$). Let further $E_{\ell\nu}$ denote the finite dimensional representation contained in $\chi^{-}_{\ell\nu}$ (as well as the space in which it acts). Then the character $\Theta_{\ell\nu}$ of $U_{\ell\nu}$ can be expressed in terms of the character $\Theta_{E_{\ell\nu}}$ of $E_{\ell\nu}$ and characters of elementary representations (see Sec. 3B) as follows:

$$\Theta_{\ell\nu} = \Theta_{\chi'^{+}_{\ell\nu}} - \Theta_{\chi^{+}_{\ell\nu}} + \Theta_{E_{\ell\nu}} . \tag{7.4}$$

7.B Unitarily induced representations on G/K

To identify the discrete series representations, we use a method which is fairly standard and has been applied to arbitrary semisimple Lie groups [H7, S2']. It singles out the discrete series IR's as subrepresentations of suitable unitarily induced representations of G [$= SO^{\uparrow}(2n,1)$] on G/K.

Let $\tau = \tau(k)$ be a unitary IR of K on a finite dimensional vector space $\mathcal{V}_\tau$. Consider the space $\mathcal{L}^2_\tau = \mathcal{L}^2(dg, \mathcal{V}_\tau)$ of functions f on G with values in $\mathcal{V}_\tau$ which satisfy the covariance property

$$\mathfrak{f}(gk) = \tau(k^{-1})\mathfrak{f}(g) \tag{7.5}$$

and the square integrability condition

$$(\mathfrak{f},\mathfrak{f}) = \int_G \langle \mathfrak{f}(g), \mathfrak{f}(g) \rangle dg < \infty \tag{7.6}$$

Here $\langle\ ,\ \rangle_\tau$ is the hermitian (sesquilinear) scalar product in $\mathcal{V}_\tau$ and dg is (as usual) the Haar measure on G. We define a (highly reducible) unitary representation U_τ of G on $\mathcal{L}^2_\tau$ by

$$[U_\tau(g)\mathfrak{f}](g') = \mathfrak{f}(g^{-1}g'). \tag{7.7}$$

U_τ can be identified with a subrepresentation of the left regular representation of G ($\mathcal{L}^2_\tau$ is isomorphic to a closed subspace of the corresponding representation space). Therefore, it admits a decomposition into irreducibles in which only principal series and discrete series representations of G appear (see Sec. 8 below).

The procedure of picking a single discrete series representation out of U_τ is based on the following two observations.

(i) A standard reciprocity theorem (similar to the one quoted in Sec. 2C) says that U_τ contains the discrete series representation U of G iff the restriction of U to K contains τ.

(ii) For every IR of G the Casimir operator $\mathcal{C}_2$ must be a multiple of the identity: $\mathcal{C}_2\mathfrak{f} = \lambda\mathfrak{f}$. The eigenfunctions of $\mathcal{C}_2$ belong to $\mathcal{L}^2_\tau$ iff the corresponding eigenvalue λ is discrete. The representation $U_{\tau\lambda}$ in the corresponding eigensubspace $\mathcal{L}^2_{\tau\lambda}$ of $\mathcal{L}^2_\tau$ is square integrable.

Consider now those groups $SO^{\uparrow}(2n,1)$, for which K is semisimple (this excludes $SO^{\uparrow}(2,1)$). According to Corollary 7.5 (Sec. 7A) only the group $G = SO^{\uparrow}(4,1)$ may have discrete series of type I unitary representations. On the other hand Eq. (6.8) gives us the K-content of the possible candidates U for the discrete series representations in this case. A fixed eigenvalue λ of $\mathcal{C}_2$ may correspond, in general, to a finite number of inequivalent discrete series representations $U_{\lambda i}$ (there are two such representations for each eigenvalue λ in the case of the group G under consideration). For every given $U = U_{\lambda i_0}$ we select a unitary IR τ of K which is contained in $U_{\lambda i_0}$, but does not appear in $U_{\lambda i}$ for $i \neq i_0$. This guarantees that the subspace $\mathcal{L}^2_{\tau\lambda} \subset \mathcal{L}^2_\tau$ which consists of all functions $f \in \mathcal{L}^2_\tau$ satisfying the differential equation $(\mathcal{C}_2 - \lambda) f = 0$ carries an irreducible discrete series representation, equivalent to U.

It follows from (6.8) and from the explicit expression (3.35) for the eigenvalues of $\mathcal{C}_2$ that a suitable choice for τ belonging to the representation $U^{\pm}_{\ell\nu}$, which acts in $D^{\pm}_{\ell\nu}$, is

$$\tau_s^+ = [s, s] \subset U_{\ell\nu}^+, \quad \tau_s^- = [-s, s] \subset U_{\ell\nu}^- \qquad \text{for } s = \ell+\nu. \tag{7.8}$$

It coincides with the choice made by Takahashi [T1], but it is not unique. Another choice is made in ref. [H7],

We shall consider the case $\tau = \tau_s^+$ and will write $U_\tau = U_s^+$. The case $\tau = \tau_s^-$ is quite similar (and can be obtained from the first one by taking the mirror image).

7.C Realization of the unitary representation U_s^+ in the space $\mathcal{L}^2_{\alpha_{s+}}(\tilde{N}A)$

The covariance property (7.5) implies that the functions $f(g) \in \mathcal{L}^2_\tau$ are determined from their values on the homogeneous space G/K which is isomorphic as a manifold to the subgroup $\tilde{N}A$ of G (cf. the Iwasawa decomposition (1.12)). For $G = SO^\uparrow(4,1)$ that is the 4-dimensional manifold

$$B \equiv G/K = SO^\uparrow(4,1)/SO(4) \simeq \tilde{N}A \simeq \mathbb{R}^3 \times \mathbb{R}^1_+ . \tag{7.9}$$

It can be parametrized by 4-vectors

$$\vec{x} = (x, y), \; x = (x_1, x_2, x_3) \in \mathbb{R}^3, \quad y > 0, \tag{7.10}$$

or by the corresponding matrices $b = \tilde{n}a$ in $\tilde{N}A$:

$$b_{\vec{x}} = \tilde{n}_x a_y = \begin{pmatrix} \mathbb{1} & -x/y & x/y \\ x & \frac{y^2 - x^2 + 1}{2y} & \frac{\vec{x}^2 - 1}{2y} \\ x & \frac{y^2 - x^2 - 1}{2y} & \frac{\vec{x}^2 + 1}{2y} \end{pmatrix}, \quad (y = |a_y|, \; \vec{x}^2 = x^2 + y^2). \tag{7.11}$$

c) <u>Special compact transformations</u> (rotations in the (3, 4)-plane):

Let

$$k_t = \begin{pmatrix} 1 & 0 & 0 & 0 & 0 \\ 0 & 1 & 0 & 0 & 0 \\ 0 & 0 & \frac{1-t^2}{1+t^2} & -\frac{2t}{1+t^2} & 0 \\ 0 & 0 & \frac{2t}{1+t^2} & \frac{1-t^2}{1+t^2} & 0 \\ 0 & 0 & 0 & 0 & 1 \end{pmatrix} \tag{7.19}$$

then

$$k_t^{-1} b_{\vec{x}} = b_{\vec{x}'(t)}\, k(-t; \vec{x}), \tag{7.20a}$$

where

$$x_i' = \frac{1+t^2}{\sigma}\, x_i,\ i = 1, 2,\ x_3' = \frac{(1-t^2)x_3 + t(\vec{x}^2 - 1)}{\sigma},\quad y' = \frac{1+t^2}{\sigma}\, y\,, \tag{7.20b}$$

$$\sigma = \sigma(\vec{x}, t) = 1 + 2t x_3 + t^2\, \vec{x}^2\,,$$

$$k(-t;\vec{x}) = \left(\begin{array}{c|ccc} \delta_{ij} - \frac{2t^2}{\sigma} x_i x_j & -2\frac{t}{\sigma}(1+tx_3)x_i & -2\frac{t^2}{\sigma} y x_i & 0 \\ \hline 2\frac{t}{\sigma}(1+tx_3)x_j & 1-2\frac{t^2}{\sigma}(\vec{x}^2 - x_3^2) & 2\frac{ty}{\sigma}(1+tx_3) & 0 \\ -2t^2 \frac{yx_j}{\sigma} & -2\frac{ty}{\sigma}(1+tx_3) & 1-2\frac{t^2}{\sigma}y^2 & 0 \\ 0 & 0 & 0 & 1 \end{array}\right) = \tag{7.20c}$$

$$= m_1 \, k_{-t_1} \, m_1 , \quad t_1 = t_1(t,\vec{x}) = \frac{ty}{A} , \quad A = (1 + 2tx_3 + t^2 x^2)^{\frac{1}{2}}$$

$$m_1 = m_1(t,x) = \left(\begin{array}{c|c|c} \delta_{ij} - \frac{t^2 x_i x_j}{A(A+1+tx_3)} & -\frac{tx_i}{A} & 0 \\ \hline t\frac{x_j}{A} & \frac{1+tx_3}{A} & 0 \\ \hline 0 & 0 & \mathbb{1}_2 \end{array}\right) \qquad i,j = 1,2; \tag{7.20d}$$

finally, if we use the same notation m_1 for the non-trivial 3 x 3 sub-matrix of the above

$$\Lambda\,[m_1 \, k_{-t_1} \, m_1] = m_1 \, \Lambda\,[k_{-t_1}] \, m_1 , \tag{7.21a}$$

where k_{-t_1} is given by (7.19) and

$$\Lambda[k_{-t_1}] = \begin{pmatrix} \frac{1-t_1^2}{1+t_1^2} & \frac{2t_1}{1+t_1^2} & 0 \\ \frac{-2t_1}{1+t_1^2} & \frac{1-t_1^2}{1+t_1^2} & 0 \\ 0 & 0 & 1 \end{pmatrix} \qquad (7.21b)$$

We notice that the conformal inversion R (1.29) is a proper $SO^{\uparrow}(4,1)$ transformation (corresponding to simultaneous rotation in π in the (1,2) and (3,4) planes) such that

$$R\vec{x} = \left(-\frac{x}{\vec{x}^2}, \frac{y}{\vec{x}^2}\right), \quad k(R,\vec{x}) = \begin{pmatrix} \frac{2x_\mu x_\nu}{\vec{x}^2} - \delta_{\mu\nu} & \frac{2x_\mu y}{\vec{x}^2} & 0 \\ -2\frac{yx_\mu}{\vec{x}^2} & 1-2\frac{y^2}{\vec{x}^2} & 0 \\ 0 & 0 & 1 \end{pmatrix} \qquad (7.22a)$$

$$(\Lambda[k(r,\vec{x})])_{\mu\nu} = \frac{1}{\vec{x}^2}\left[2(x_\mu x_\nu - y\,\varepsilon_{\mu\nu\lambda}\,x_\lambda) + (y^2 - x^2)\,\delta_{\mu\nu}\right], \mu, \nu = 1,2,3 \qquad (7.22b)$$

7.D K - invariants. Solution of the eigenvalue problem for the Casimir operator. The discrete series $U^+_{\ell\nu}$

The infinitesimal generators of the representation U_τ constructed in the previous subsection can be written down as first order differential operators in $\vec{x}$ and y, in a similar way as the generators of the elementary representations displayed in Sec. 3D. In particular, the generators of ($\mathbb{R}^3$) translations and M = SO(3) rotations are given by exactly the same formulas (3.29-31) (with 2h=3). The generators of dilatations and special conformal transformations include in the present case the variable y:

$$D(= X_{40}) = - \vec{x}\vec{\nabla} \equiv - x\nabla_x - y\nabla_y \tag{7.23}$$

$$C_\mu (= X_{\mu 0} - X_{\mu 4}) = 2x_\mu (\vec{x}\vec{\nabla}) - \vec{x}^2 \nabla_\mu + 2x_\sigma s_{\sigma\mu} - y\varepsilon_{\mu\sigma\lambda} s_{\sigma\lambda} \tag{7.24}$$

$$(\mu = 1, 2, 3),$$

where $s_{\mu\nu}$ is the spinorial part of the rotation generators, given by (3.31). The Casimir operator (3.34) is a second order differential operator:

$$\mathcal{C}_2(U_\tau) = \tfrac{1}{2} X_{AB} X^{BA} = y^2 \vec{\nabla}^2 - 2y\nabla_y + s(s+1) + 2y(s_{23}\nabla_1 + s_{31}\nabla_2 + s_{12}\nabla_3). \tag{7.25}$$

Here we have used that

$$\tfrac{1}{2} s_{\mu\lambda} s^{\lambda\mu} = s(s+1) \tag{7.26}$$

for the representation $\tau_s = (s, s)$ of K = SO(4) (which reduces to the representation (s) of SO(3) when restricted to M).

In order to find the discrete spectrum of $\mathcal{C}_2$ it is sufficient to study some special class of solutions of the eigenvalue equation

$$\mathcal{C}_2 \Phi = [s(s+1) + (\ell - 1)(\ell + 2)]\Phi \; . \qquad (7.27)$$

[In writing the eigenvalues of $\mathcal{C}_2$ in such a special form, we have anticipated the result, given by Eq. (3.35) with ℓ replaced here by s, $c = h + \ell - 1$, $h = \frac{3}{2}$. However, we need not assume for the moment that ℓ is a non-negative integer (or even that it is real). This will come out as a result of the solution of the eigenvalue problem.]

We shall look for solutions in which the $\mathfrak{z}$-dependence is factored out in the following simple form:

$$\phi(\vec{x}; \mathfrak{z}) = \phi_{s\ell}(\vec{x}) (\mathfrak{z} \overset{+}{Q} (x, y+1) \mathfrak{z}_{\phi})^{s} , \qquad (7.28)$$

where

$$(\mathfrak{z} \overset{+}{Q}(x, y+1) \mathfrak{z}_{\phi}) = \mathfrak{z}^{\lambda} \overset{+}{Q}_{\lambda\mu}(x, y+1) \mathfrak{z}^{\mu}_{\phi} \qquad (7.29)$$

will be determined by the requirement of K-covariance, $\mathfrak{z}_{\phi}$ is a fixed vector of $\mathbb{K}_3$ (i.e. $\mathfrak{z}^2_{\phi} = 0$, and different $\mathfrak{z}_{\phi}$ will correspond to different solutions ϕ of Eq. (7.27)). (The reason for the peculiar way of writing the second argument of $\overset{+}{Q}$ will become clear in the next subsection.) We shall further assume that $\phi_{s\ell}(\vec{x})$ is K invariant (which does not impose any restriction on the eigenvalues). Under these assumptions the vector (7.28) belongs to the K-irreducible subspace of the representation $\tau_s = (s, s)$ of SO(4). It belongs to the class of <u>K-finite vectors</u>

which plays an important role in the algebraic approach to representation theory. The above choice will not only allow us to solve easily the eigenvalue problem, but will also prepare the ground for finding the G-invariant 2-point Green function associated with Eq. (7.27) (see Sec. 7E below).

We start by writing down the most general K-invariant function of $\vec{x}$. There is just one algebraic SO(4) invariant, which will be determined as follows. We first observe that every $G = SO^{\uparrow}(4,1)$ invariant of the two points $\vec{x}_1$ and $\vec{x}_2$ is a function of

$$u = u(\vec{x}_1; \vec{x}_2) = \frac{4y_1y_2}{(x_1-x_2)^2 + (y_1+y_2)^2} \tag{7.30}$$

[One way to establish that is to take a homogeneous rational function of y_1, y_2 and $(x_1-x_2)^2$, which is automatically $\overline{N}AM$ invariant, and to impose invariance under the conformal inversion R (7.22a).] Then, we notice, that the point $\vec{x} = (0,1)$ is K-invariant (since $b_{(0,1)} = 1$), and deduce that

$$u = u(\vec{x}; 0,1) = \frac{4y}{x^2 + (y+1)^2} \qquad (0 < u \leq 1) \tag{7.31}$$

is a basic 1-point K-invariant (every K-invariant of $\vec{x}$ is a function of u).

In order to find Q^+ we demand that

$$\mathfrak{z}_1 Q^+(0, y_+ + 1)\mathfrak{z}_2 = \mathfrak{z}_1\mathfrak{z}_2 \tag{7.32}$$

for

$$y_+ = u^{-1} (1 + \sqrt{1-u})^2 \quad > 1 \qquad (0 < u < 1). \tag{7.33}$$

Then $\mathfrak{z}_1 \overset{+}{Q}(x, y+1) \mathfrak{z}_2$ is obtained by applying the (compact) "boost"

$$k_{\vec{x}} = m_x \; k_{-t(\vec{x})} \; m_x^{-1} \;, \qquad (0, y_+)' = (x, y) \tag{7.34a}$$

where $k_{-t(\vec{x})}$ is given by (7.19) with

$$t(\vec{x}) = \frac{\sqrt{[x^2+(y+1)^2][x^2+(y-1)^2]} + 1 - x^2 - y^2}{2\sqrt{x^2}} \;, \quad \frac{4y}{x^2+(y+1)^2} = u \tag{7.34b}$$

The non trivial 3 x 3 part of m_x is given by

$$m_x = \left(\begin{array}{c|c} \delta_{ij} - \dfrac{\hat{x}_i \hat{x}_j}{1+\hat{x}_3} & \hat{x}_i \\ \hline \hat{x}_j & \hat{x}_3 \end{array} \right), \hat{x} = \frac{x}{\sqrt{x^2}} \;, \quad i, j = 1, 2. \tag{7.34c}$$

Using further the explicit form (7.21) of the representation $\Lambda[k]$ of SO(4) we find

$$(\mathfrak{z} \overset{+}{Q}(x, y+1) \mathfrak{z}_\phi) = (\Lambda[k_{\vec{x}}]\mathfrak{z} \;, \mathfrak{z}_\phi) = \frac{u}{2y} [(x\mathfrak{z})(x\mathfrak{z}_\phi) + (y+1)^2 \mathfrak{z}\mathfrak{z}_\phi +$$

$$+ (y+1)(x\, \mathfrak{z} \wedge \mathfrak{z}_\phi)] - \mathfrak{z}\mathfrak{z}_\phi =$$

$$= \frac{2(x\mathfrak{z})(x\mathfrak{z}_\phi) + [(y+1)^2 - x^2]\mathfrak{z}\mathfrak{z}_\phi + (y+1)(x\,\mathfrak{z} \wedge \mathfrak{z}_\phi)}{x^2 + (y+1)^2} \tag{7.35}$$

where $(\mathfrak{z} \wedge \mathfrak{z}_\phi)_\lambda = \varepsilon_{\lambda\mu\rho} \mathfrak{z}_\mu \mathfrak{z}_{\phi\rho}$.

Inserting (7.28) (with $\phi_{s\ell} = \phi_{s\ell}(u)$) in (7.27) and taking into account the differentiation properties

$$\frac{\partial u}{\partial x_\mu} = -\frac{u^2}{2y} x_\mu, \quad y\frac{\partial u}{\partial y} = u(1 - \frac{y+1}{2} u),$$

$$(y^2 \vec{\nabla}^2 - 2y \nabla_y)u = -2u ; \tag{7.36a}$$

$$\{y^2 \vec{\nabla}^2 - 2y\nabla_y + 2y (\nabla_1 s_{23} + \nabla_2 s_{31} + \nabla_3 s_{12})\} Q^+ = 2u Q^+ \tag{7.36b}$$

we obtain the following ordinary differential equation for $\phi_{s\ell}(u)$:

$$\{ (u-1)u^2 \frac{d^2}{du^2} + 2u \frac{d}{du} + [(\ell-1)(\ell+2) - s(s+1)u] \} \phi_{s\ell}(u) = 0. \tag{7.37}$$

The solution of this equation is

$$\phi_{s\ell}(u) = cu^{\ell+2} F(\ell + s + 2, \ell + 1 - s; 2\ell + 2; u) \tag{7.38}$$

(see [K1]). It is a polynomial in u (square integrable with respect to the scalar product (7.13) with volume element $y^{-4} d\vec{x}$) if ℓ is a non negative integer smaller than s:

$$s - \ell = \nu \ (=1, 2, \ldots), \quad \ell = 0, 1, 2, \ldots \tag{7.39}$$

We see that the eigenfunctions (7.28) (7.35) (7.36) are (real) analytic in the half space $y > 0$ (in other words they belong to $\mathcal{A}_\tau$). In addition, the functions $y^{-\ell-2} \phi(\vec{x}; \mathfrak{z})$ have a C^∞ limit for $y \to 0$ (which will be shown in the next subsection to belong to $D^+_{\ell\nu}$). The set of all real analytic solutions of (7.27) which belong to $\mathcal{H}_\tau$ and have the above properties will be denoted by $\mathcal{A}_{\ell\nu}$

Thus we define the (Hilbert) subspace $\mathcal{H}^{+}_{\ell\nu}$ of $\mathcal{H}_{\tau}$ as the set of (square integrable) solutions of (7.27) with s and ℓ satisfying (7.39), and identify the discrete series representation $U_{\ell\nu}$ with the restriction of U_{τ} on $\mathcal{H}^{+}_{\ell\nu}$. Its mirror image $U^{-}_{\ell\nu}$ can be defined as the IR obtained in the same way from $U_{\bar{\tau}}$, where

$$\bar{\tau}_s = [-s, s] \qquad s = \ell + \nu. \tag{7.40}$$

As it was mentioned before the representation

$$U_{\ell\nu} = U^{+}_{\ell\nu} + U^{-}_{\ell\nu} \tag{7.41}$$

is irreducible with respect to the extended group $O^{\uparrow}(4,1)$. Similarly, we shall write for the (dense) subspaces of real analytic functions

$$\mathcal{A}_{\ell\nu} = \mathcal{A}^{+}_{\ell\nu} \oplus \mathcal{A}^{-}_{\ell\nu} \subset \mathcal{H}_{\ell\nu} = \mathcal{H}^{+}_{\ell\nu} + \mathcal{H}^{-}_{\ell\nu} .$$

The functions (7.28)(7.35)(7.38) are finite linear combinations of the vectors of the canonical basis [defined in Sec. 2C(see also Appendix A)]. If we act on them by (some of) the non-compact generators [say D(7.23)] we will obtain the basis vectors for "higher" representations of K. In this way one can verify that the representations $U^{\pm}_{\ell\nu}$ have the same K-content (6.8) as the elementary subrepresentations acting in $D^{\pm}_{\ell\nu}$.

7.E Two-point Green function. Equivalence of $U^{+}_{\ell\nu}$ with the sub-representation of $\chi'^{+}_{\ell\nu}$ acting in $D^{+}_{\ell\nu}$

The functions $\phi(\vec{x})$ of $\mathcal{A}^{+}_{\ell\nu}$ (which behave like $y^{\ell+2} f(x)$ for $y \to 0$) are in one-to-one correspondence with the elements $f(x)$ of $D^{+}_{\ell\nu}$. The mapping $L: \mathcal{A}^{+}_{\ell\nu} \to D^{+}_{\ell\nu}$ is given by

$$f(x) = [L\phi](x) = \lim_{y \downarrow 0} y^{-\ell-2} \phi(x,y). \tag{7.42}$$

Let $T^{+}_{\ell\nu}$ be the restriction on $D^{+}_{\ell\nu}$ of the operators T^{χ} of the elementary representation $\chi'^{+}_{\ell\nu}$. It is not difficult to verify that L plays the role of an intertwining operator between the representation $U^{+}_{\ell\nu}$ (on $\mathcal{A}^{+}_{\ell\nu}$) and $T^{+}_{\ell\nu}$:

$$L\, U^{+}_{\ell\nu} = T^{+}_{\ell\nu}\, L\,. \tag{7.43}$$

In order to establish the equivalence between $U^{+}_{\ell\nu}$ and $T^{+}_{\ell\nu}$ we have to construct the inverse operator L^{-1}. This is done in a standard fashion in terms of the Green function $\mathcal{D}^{+}$ of the operator $\mathcal{C}_2 - \lambda$ $(\lambda = (\nu+\ell)(\nu+\ell+1) + (\ell-1)(\ell+2))$.

The G-invariant solution of the equation

$$(\mathcal{C}_2 - \lambda)\mathcal{D}^{+} \equiv [y^2\vec{\nabla}^2 - 2y\nabla_y + 2y(s_{23}\nabla_1 + s_{31}\nabla_2 + s_{12}\nabla_3) - (\ell-1)(\ell+2)] \cdot$$

$$\cdot\, \mathcal{D}^{+}(\vec{x}, \mathfrak{z}; \vec{x}', \mathfrak{z}') = y^4\, \delta(\vec{x}-\vec{x}')(\mathfrak{z}\mathfrak{z}')^{\ell+\nu} \tag{7.44}$$

is

$$\mathcal{D}^{+}(\vec{x}, \mathfrak{z}; \vec{x}', \mathfrak{z}') = N_{\ell\nu}\, u^{\ell+2}\, F_{\ell\nu}(u)\, [\mathfrak{z}\, Q^{+}(x-x', y+y')\, \mathfrak{z}']^{\ell+\nu} \tag{7.45}$$

where $u = u(\vec{x}, \vec{x}')$ is the invariant variable (7.30), Q^+ is given by (7.35) and $F_{\ell\nu}$ is the singular solution of the hypergeometric equation

$$\{u(1-u) \frac{d^2}{du^2} + 2[\ell + 1 - (\ell + 2)u] \frac{d}{du} + (\nu-1)(2\ell+\nu+2)\} F_{\ell\nu}(u) = 0 \tag{7.46a}$$

(given by Eq. 9.153.3 of ref. [G7]). We will only need to know the behavior of $F_{\ell\nu}$ in the neighborhood of the singular points $u = 0$ and $u = 1$; it is

$$F_{\ell\nu}(u) \approx - \left[\binom{2\ell + \nu + 1}{\nu} \binom{2\ell + \nu}{\nu - 1}\right]^{-1} u^{-2\ell-1} \quad \text{for } u \to 0 \tag{7.46b}$$

$$F_{\ell\nu}(u) \approx (-1)^{\nu-1} \frac{(\nu-1)!}{(2\ell+2)_\nu} \frac{1}{1-u} \quad \text{for } u \to 1. \tag{7.46c}$$

The singularity for $u \to 1$ is responsible for the δ-function in the right-hand side of (7.44). The overall normalization $N_{\ell\nu}$ is determined from the Green formula (for ϕ satisfying (7.27))

$$\phi(\vec{x}, z) = \frac{1}{(\ell+\nu)!} \int_{B_{\vec{x}}} \{[\mathcal{C}_2' \mathcal{D}^+(\vec{x}, z; \vec{x}', \partial_{z'})] \phi(\vec{x}', z') - \mathcal{D}^+(\vec{x}, z; \vec{x}', \partial_{z'}) [\mathcal{C}_2' \phi(\vec{x}', z')]\} \frac{d^3x dy}{y^4}$$

$$= - \frac{1}{(\ell+\nu)!} \int_S \{[\frac{\partial}{\partial n'} \mathcal{D}^+(\vec{x}, z; \vec{x}', \partial_{z'})] \phi(\vec{x}', z') - \mathcal{D}^+(\vec{x}, z; \vec{x}', \partial_{z'}) \frac{\partial}{\partial n'} \phi(\vec{x}', z')\} \frac{d\sigma}{y^2},$$

where $B_{\vec{x}}$ is chosen as a small ball in $\mathbb{R}^4$ with center in $\vec{x}$, S is its surface (with surface element $d\sigma$), the prime on $\mathcal{C}_2'$ and $\frac{\partial}{\partial n'}$ indicates that the differential operators act on the primed argument and the normal n' points inside $B_{\vec{x}}$. The result is

$$N_{\ell\nu} = \frac{(-1)^{\nu}}{(4\pi)^2} \frac{(2\ell+2)_{\nu}}{(\nu-1)!} . \tag{7.47}$$

Applying the same Green formula to the entire half space $\mathbb{R}^3 \times \mathbb{R}_+$, we express $\phi(\vec{x})$ in terms of its boundary value f(x) (7.42). This gives us an explicit construction of the operator L^{-1}. Only the term (7.46b) of $F_{\ell\nu}$ enters the final formula; we have

$$\phi(\vec{x}, \mathfrak{z}) = [L^{-1}f](\vec{x}, \mathfrak{z}) = \frac{1}{(\ell+\nu)!} \int S^+_{\ell\nu}(\vec{x}, \mathfrak{z}; x', \partial_{\mathfrak{z}'}) f(x', \mathfrak{z}')d^3x' \tag{7.48}$$

where $f \in D^+_{\ell\nu}$ and

$$S^+_{\ell\nu}(\vec{x}, \mathfrak{z}; x', \partial_{\mathfrak{z}'}) = - \lim_{y' \downarrow 0} [y'^{\ell} \frac{\partial}{\partial y'} \mathcal{D}^+(\vec{x}, \mathfrak{z}; \vec{x}', \partial_{\mathfrak{z}'}) -$$

$$- (\ell + 2)y'^{\ell-1} \mathcal{D}^+(\vec{x}, \mathfrak{z}; \vec{x}', \partial_{\mathfrak{z}'})]$$

$$= \frac{(-1)^{\nu-1}}{(4\pi)^2} \frac{\nu!(2\ell+1)!}{(2\ell+\nu)!} \left[\frac{4y}{(x-x')^2+y^2}\right]^{1-\ell} [\mathfrak{z}\, Q^+(x-x', y)\partial_{\mathfrak{z}'}]^{\ell+\nu} . \tag{7.49}$$

A similar expression is obtained for $\phi(\vec{x}) \in \mathcal{A}^-_{\ell\nu}$ with $f \in D^-_{\ell\nu}$ and Q^+ replaced by Q^- where

$$Q^{\pm}_{\mu\rho}(x,y) = \frac{2x_\mu x_\rho + (y^2 - x^2)\,\delta_{\mu\rho} \pm y x_\lambda \varepsilon_{\lambda\mu\rho}}{x^2 + y^2}\ , \quad \mu,\rho = 1,2,3\ . \tag{7.50}$$

Because of the orthogonality of the two spaces

$$\frac{1}{(\ell+\nu)!} \int S^-_{\ell\nu}(\vec{x}, \mathfrak{z}; x', \partial_{\mathfrak{z}'})\, f(x', \mathfrak{z}')\, d^3x' = 0 \quad \text{for } f \in D^+_{\ell\nu} \tag{7.51}$$

where $S^-_{\ell\nu}$ is given by the counterpart of (7.49) with Q^+ replaced by Q^-.

Thus we established the equivalence between the representations $U^+_{\ell\nu}$ and $T^+_{\ell\nu}$. It implies the proportionality of the scalar products (7.13) and (6.26-27) in the corresponding representation space. It seems rather intricate to derive this proportionality directly from the definition of the two scalar products. In order to show what is involved, we shall verify the above statement in the simplest case $\ell = 0$ $\nu = 1$. (In Appendix B.4 we present a complete discussion of this point for the analytic discrete series of $SO^{\uparrow}(2,1) \approx SL(2,\mathbb{R})/Z_2$.)

Proposition 7.6. The elements of D^+_{01} are 3-vector functions of C'^+_{01} which satisfy the identity

$$f(x) = \nabla \wedge (-\Delta)^{-\frac{1}{2}} f(x) = \mathrm{curl}_x \int \frac{f(x')}{(x-x')^2} \frac{dx'}{(2\pi)^2}\ . \tag{7.52}$$

For functions in D_{01} the hermitian form $(f_1, B'^+_{01} f_2)$ (6.26) (6.27) can be written in the following way:

$$(f_1, B'^{+}_{01} f_2) = \frac{1}{\sqrt{2}\,\pi^2} \iint \frac{\bar{f}_1(x_1) f_2(x_2)}{x_{12}^2}\, dx_1\, dx_2 \quad (\bar{f}_1 f_2 \equiv \bar{f}_{1\mu} f_{2\mu}) \tag{7.53}$$

Proof. According to the remark following Eq. (5.20) the projection operator $\Pi(p)$ $(= \Pi^{11}(p))$ can be split into two (covariant) one dimensional projectors (for $2h = 3$):

$$\Pi(p) = \Pi^{+}(p) + \Pi^{-}(p). \tag{7.54a}$$

where

$$\Pi^{\pm}_{\mu\nu}(p) = \tfrac{1}{2}(\delta_{\mu\nu} - \hat{p}_\mu \hat{p}_\nu \pm i\hat{p}_\lambda \varepsilon_{\mu\lambda\nu}) \qquad (\hat{p} = (p^2)^{-\frac{1}{2}}\, p)\,. \tag{7.54b}$$

Π^{+} projects on the invariant subspace D^{+}_{01} of D_{01}. Eq. (7.52) expresses the fact that any $f \in D^{+}_{01}$ has zero divergence and is orthogonal to Π^{-}:

$$(\nabla f)(x) = 0 \qquad (\forall\, f \in D_{01}) \tag{7.55a}$$

$$(\Pi^{-} f)(x) = 0 \qquad (f \in D^{+}_{01}) \tag{7.55b}$$

In order to rewrite the scalar product (6.26) (6.27) in the simple form (7.53), we use the identities

$$\frac{1}{x_{12}^2}\, r_{\mu\nu}(x_{12}) = -\nabla_{1\mu} \frac{(x_{12})_\nu}{x_{12}^2}\,, \qquad \frac{2(x_{12})_\mu (x_{12})_\nu}{x_{12}^4} = \frac{\delta_{\mu\nu}}{x_{12}^2} + \tfrac{1}{2}\nabla_{1\mu}\nabla_{2\nu} \ln x_{12}^2\,,$$

and integrate by parts taking into account (7.55a).

Now we shall demonstrate that the scalar product (7.13) for functions in $\mathcal{A}^+_{01}$ can also be reduced to the form (7.53).

According to (7.48-51) the function

$$f(x, y) = y^{-2}\,\phi(x, y), \qquad (\phi \in \mathcal{A}^+_{01}) \tag{7.56}$$

is expressed in terms of its boundary value $f(x, 0) = f(x)$ by

$$f(x, y) = \frac{1}{(2\pi)^2}\,\nabla_x \wedge \int \frac{f(x')\,dx'}{(x-x')^2 + y^2} = \frac{y}{\pi^2}\int \frac{f(x')\,dx'}{[(x-x')^2 + y^2]^2}\,. \tag{7.57}$$

The scalar product (7.13) is expressed in the following way in terms of the functions $f(x') \in D^+_{01}$ (using the second equality (7.57)):

$$(\phi_1, \phi_2) = \int \bar{f}_1(x, y)\, f_2(x, y)\, dx\, dy =$$

$$= \frac{1}{(2\pi)^2} \iint \frac{\bar{f}_1(x_1)\, f_2(x_2)}{x_{12}^2}\, dx_1\, dx_2 = 2^{-\frac{3}{2}}\,(f_1, B'^{+}_{01}\, f_2)\,. \tag{7.58}$$

8. The Plancherel theorem. Concluding remarks

8.A Harmonic analysis of the left regular representation of $SO^{\uparrow}(2h+1, 1)$ for integer h

Here we shall apply our knowledge of the IR's to the standard problem of harmonic analysis of functions on G.

Let $D = D(G)$ be the space of all infinitely differentiable scalar functions $\varphi(g)$ on G of compact support. Let $\hat{G}$ be the set of all (equivalence classes of) unitary IR's of G and let $\kappa \in \hat{G}$ be realized by operators $T_\kappa(g)$ acting in a Hilbert space $\mathcal{H}_\kappa$. Let $\Theta_\kappa(\varphi)$ be the character (3.7) of the representation κ. The Plancherel problem is to find a positive measure $d\kappa$ on $\hat{G}$ such that

$$\varphi(1) = \int_{\hat{G}} \Theta_\kappa(\varphi)\, d\kappa\,, \quad \text{for all } \varphi \in D\,. \tag{8.1}$$

where 1 is the unit element of G.

For $\varphi \in D$ define $\varphi_g \in D$ in terms of the left regular action of the group:

$$\varphi_g(g') = [L_{g^{-1}}\varphi](g') = \varphi(gg')\,. \tag{8.2}$$

Then Eq. (8.1) implies that

$$\varphi(g) = \int_{\hat{G}} \operatorname{Tr} T_\kappa(\varphi_g)\, d\kappa = \int_{\hat{G}} \operatorname{Tr}(T_\kappa^*(g)\, T_\kappa(\varphi))\, d\kappa \tag{8.3}$$

(we have used that T_κ is unitary, so that $T_\kappa(g^{-1}) = T_\kappa^*(g)$).

Multiplying both sides of Eq. (8.3) by $\overline{\varphi(g)}$ and integrating over $g \in G$, we obtain

$$\int |\varphi(g)|^2 \, dg = \int \mathrm{Tr}\,(T_\kappa(\varphi)^* \, T_\kappa(\varphi) \, d\kappa \; . \tag{8.4}$$

Using continuity of both sides with respect to the Hilbert space topology in $\mathcal{L}^2(G)$ we can extend the validity of Eq. (8.4) to all square integrable functions φ on G.

In order to solve the Plancherel problem one has first to know which unitary IR's enter into the integral in the right hand side of (8.1). A partial solution of the problem for the generalized Lorentz group is given by

Theorem 8.1.([N2], [H5]) Let $G = SO^{\uparrow}(2h+1, 1)$ (or Spin(2h+1,1)) with integer h. Then only the principal series of unitary IR's $\chi = [(\ell), c(=i\sigma)]$ enter into the Plancherel formula (8.1) (8.4) and

$$\int d\kappa = \sum_{(\ell)\in \hat{M}} \int_{c=-i\infty}^{\infty} \rho(\ell, c) \frac{dc}{2\pi i} \tag{8.5}$$

with the Plancherel measure ρ expressed most easily in terms of the variables n_j, defined in (2.3):

$$\rho(\ell, c) = C \prod_{1\leq i< j\leq h} (n_j^2 - n_i^2) \prod_{k=1}^{h} (n_k^2 - c^2),$$

$$n_j = \ell_j + j - 1 \; (j = 1, \ldots, h), \; c = i\sigma \tag{8.6}$$

(C being a constant, depending on the normalization of the Haar measure on G) ; the sum in (8. 5) is carried over the set $\hat{M}$ of (equivalence classes of unitary) IR's of M .

We see that in the special case of type I representations the weight $\rho(\ell, i\sigma)$ is proportional to the absolute value square of the normalization factor $n(\chi)$ [(5. 24) or (5. 26)] of the intertwining operator:

$$\rho(\ell, i\sigma) = A_\ell |n(\chi)|^2 = A_\ell \left| \frac{\Gamma(h+i\sigma-1)}{\Gamma(i\sigma)} \right|^2 [(h+\ell-1)^2 + \sigma^2] , \qquad (8.7)$$

with $A_\ell = C \prod_{1 \leq i < j \leq h-1} (j-i)(j+i-2) \frac{(\ell+h-1)!}{\ell!}$.

It was proven by Knapp and Stein [K3] (see also [W2]) that such a relation is valid for all principal series representations; moreover, the analytic continuation of ρ for non-imaginary c is given by

$$\rho(\ell, c) = A_\ell n(\chi) n(\tilde{\chi}) . \qquad (8.7)$$

8.B Harmonic analysis on $SO^{\uparrow}(2n,1)$. The role of the discrete series

For the group $SO^{\uparrow}(2n, 1)$ [or Spin(2n,1)] (n integer) the principal series representations are no longer the only ones entering into the Plancherel formula. We will explain the reason for this difference. To do that we shall use the function $F_{\varphi}(ma)$ (3.8), introduced in Sec. 3B.

A function $\varphi \in D(G)$ is called a cusp form if

$$F_{\varphi}(ma) = 0 \quad \text{for all } ma \in MA, \ a \neq 1. \tag{8.8}$$

For $G = SO^{\uparrow}(2n,1)$ (or, more generally, for any semi simple Lie group G, which has a compact Cartan subalgebra) there exists a cusp form $\varphi(g)$ depending only on the conjugacy class of g, i.e. such that

$$\varphi(gg'g^{-1}) = \varphi(g'), \text{ for all } g, g' \in G, \tag{8.9}$$

which cannot be expanded in principal series representations alone. To see that, we shall introduce two more auxiliary notions: the notions of a hyperbolic and of an elliptic set of G.

Consider the torus (= abelian compact connected Lie group) $Tor^{(q)}(G)$ of elements of G of the form

$$t = \exp\{\sum_{k=1}^{q} \theta_k X_{2k-1\,2k}\} \quad (q = n, n-1). \tag{8.10}$$

We call a t singular if $\theta_j = \theta_k$ for $j \neq k$ or $\theta_j = 0$ for some j (in particular, the identity, t = 1 is singular). The singular elements of $Tor^{(q)}$ form a lower dimensional manifold. If $t \in Tor^{(q)}$ is not singular it is called regular. The set of regular elements of $Tor^{(q)}$ will be denoted by $'Tor^{(q)}$.

We shall also use the notation $'A = \{a \in A;\ a \neq 1\}$. We define the elliptic and hyperbolic sets, $G_{e\ell}$ and G_{hyp}, of G by

$$G_{e\ell} = \{ g \in G;\ \exists\ g_1 \in G,\ t \in {}'Tor^{(n)}:\ g = g_1 t\, g_1^{-1} \} \qquad (8.11a)$$

$$G_{hyp} = \{ g \in G;\ \exists\ g_1 \in G,\ h = ta \in {}'Tor^{(n-1)} \cdot {}'A:\ g = g_1 h\, g_1^{-1} \} \qquad (8.11b)$$

The intersection $G_{e\ell} \cap G_{hyp}$ is empty. It is easily verified by considering a neighborhood of the identity that $G_{e\ell}$ has the same dimension (as a manifold) as G. Harish-Chandra has proven that the set of regular elements of G, defined by

$$'G \equiv G_{e\ell} \cup G_{hyp}, \qquad (8.12)$$

covers G up to a lower dimensional submanifold.

Since $g\, G_{e\ell}\, g^{-1} = G_{e\ell}$, there are C^∞ functions φ on G with support on $G_{e\ell}$, which satisfy (8.9). They will have automatically compact support since $G_{e\ell}$ has a compact closure (because it consists of matrices of uniformly bounded norm $\| g \|^2 = tr(g^* g) \leq 2^{n+1}$).

Proposition 8.2. Let $\varphi(g) \in D$ and supp $\varphi \subset G_{e\ell}$. Then φ is a cusp form, so that $F_\varphi(ma)$ vanishes.

The statement follows from Lemma 3.1 and from the remark that the intersection $G_{e\ell} \cap G_{hyp}$ has Haar measure zero.

It follows from the second equality (3.18) that if supp $\varphi \subset G_{el}$, then $\Theta_\chi(\varphi) = 0$ for all elementary representations. Let φ be a cusp form satisfying (8.9). Then

$$T_\kappa(\varphi) = T_\kappa(g)\, T_\kappa(\varphi)\, T_\kappa(g^{-1})\ ;$$

since the representation T_κ is unitary and irreducible we can apply Schur's lemma and deduce that

$$T_\kappa(\varphi) = \lambda\, \mathbb{1}\ , \tag{8.13}$$

where $\lambda = \lambda_\kappa(\varphi)$ is a complex number. This allows us to rewrite the representation (8.3) in the form

$$\varphi(g) = \int_{\hat{G}} \lambda_\kappa(\varphi)\, \bar{\Theta}_\kappa(g)\, d\kappa\ . \tag{8.14}$$

The contribution of the principal (as well as of the complementary) series to the right hand side of (8.14) vanishes for g in the interior of G_{el}, while the left hand side is assumed not to be identically zero on such points. Thus, an expansion of $\varphi(g)$ in terms of principal (and complementary) series alone is not possible. However, such an expansion becomes possible, if we also include the discrete series representations. This was exhibited for K-induced representations in Sec. 7. The measure $\rho_s(c)$ in (7.8) is given again by (8.6), which for the case of $O^\uparrow(4,1)$ assumes the form

$$\rho_s(c) = \frac{\Gamma(\frac{3}{2}+s)}{2(2\pi)^{3/2}\, s!}\ \frac{\Gamma(c+\frac{1}{2})\,\Gamma(\frac{1}{2}-c)}{\Gamma(c)\,\Gamma(-c)}\ [(s+\tfrac{1}{2})^2 - c^2] =$$

$$= -\frac{(2s+1)!!}{2^{s+7/2}\,\pi\, s!}\ [(s+\tfrac{1}{2})^2 - c^2]\ c\ \mathrm{tg}\,\pi c \tag{8.15}$$

The formal dimension $d_{\ell\nu}$ of the representation $U_{\ell\nu}$ (in $D_{\ell\nu}$) is given by the residue of $\rho_{\ell+\nu}(c)$ for $c = \ell + \frac{1}{2}$

$$d_{\ell\nu}^{-1} = \operatorname{Res}_{c=\ell+\frac{1}{2}} \rho_{\ell+\nu}(c) = \frac{(2\ell+2\nu+1)!!}{2^{\ell+\nu+3/2}(2\pi)^2} \frac{(\ell+\frac{1}{2})(2\ell+\nu+1)_\nu}{(\ell+\nu)!} . \tag{8.16}$$

In the general $SO^{\uparrow}(2n,1)$ case, we have the following theorem due to Hirai [H5].

<u>Theorem 8.3.</u> <u>For $G = (S)O^{\uparrow}(2n,1)$ (or $Spin(2n,1)$) only the principal and the discrete series enter into the Plancherel formula. Using the notation Θ_δ for the character of the discrete series representation U_δ with formal dimension d_δ we obtain the following explicit form for (8.1) in this case:</u>

$$\varphi(1) = \sum_{(\ell)\in\hat{M}} \int_{-i\infty}^{i\infty} \frac{dc}{2\pi i} \rho(\ell,c)\, \Theta_{[\ell,c]}(\varphi) + \sum_\delta d_\delta^{-1} \Theta_\delta(\varphi) \tag{8.17a}$$

<u>with ρ and d given by</u>

$$\rho(\ell,c) = C(\ell_1+\tfrac{1}{2})\dots(\ell_{n-1}+n-3/2) \prod_{1\le i<j\le n-1} [(\ell_j+j-\tfrac{1}{2})^2 - (\ell_i+i-\tfrac{1}{2})^2]\, c\, \times$$

$$\times \prod_{k=1}^{n-1} [(\ell_k+k-\tfrac{1}{2})^2 - c^2]\, \mathrm{tg}\,\pi c; \qquad d_\delta^{-1} = \operatorname{Res}\rho(\delta) . \tag{8.17b}$$

8.C Synopsis on unitary type I representations. Summary of equivalence relations

In the preceding pages we presented a rather comprehensive discussion of the properties of elementary type I representations of $G_{ex} = O^{\uparrow}(2h+1, 1)$ (induced by symmetric traceless tensor representations of the parabolic subgroup MAN where $M = SO(2h)$). Here we shall summarize for the reader's convenience the main facts about such representations.

First, we give below a complete list of the (unequivalent) unitary type I irreducible representations of G_{ex}, thus summarizing the results of Secs. 3D, 5C, 6C and 7 C. We are using throughout the paper the notation $\chi = [\ell , c]$ for the elementary representations, where ℓ is the number of tensor indices, characterizing the IR of $M = SO(2h)$, and $-h-c$ is the (length) dimension fixing the representation (2. 31b) of the dilatation subgroup $A = SO^{\uparrow}(1,1)$. We recall that an elementary representation χ is irreducible unless $h+c$ is an integer of the type involved in Eq. (3. 5) and considered in Sec. 6A.

a) Principal series (Sec. 3D): c-pure imaginary, $(c = i\sigma)$, ℓ-arbitrary. (These representations are still unitary for arbitrary unitary IR's ℓ of M.)

b) Two classes of type I complementary series (Sec. 5C);

$$\ell = 0 , \quad -h < c < h \quad (h \geq 1), \tag{5. 29a}$$

$$\ell = 1, 2, \ldots, -h+1 < c < h-1 \quad (h > 1) . \tag{5. 29b}$$

In both cases the point $c = 0$ is excluded from the complementary series since it belongs to the principal series.

c) Two exceptional series of unitary representations (Sec. 6C):

(i) the subrepresentations (acting in the subspace $F_{0\nu}$) of the representation $\chi^{+}_{0\nu} = [0, \nu + h - 1]$; (ii) the subrepresentations (acting in the subspace $D_{\ell\nu}$) of the elementary representations $\chi'^{+}_{\ell\nu} = [\ell + \nu, \ell + h - 1]$. In the special case of $h = 3/2$ these latter representations are reducible under the identity component $G = SO^{\uparrow}(4, 1)$ of G_{ex} and split into two discrete series representations $U^{\pm}_{\ell\nu}$ of G (Sec. 7). For $n > 2$, the discrete series representations of $SO^{\uparrow}(2n, 1)$ are not of type I.

Secondly, we shall summarize the results about equivalences among elementary (sub) representations.

The elementary representations

$$\chi = [\ell , c] \quad \text{and} \quad \tilde{\chi} = [\tilde{\ell} , -c] \tag{8.18}$$

(where $\tilde{\ell}$ is the mirror image of ℓ defined in Sec. 2A) are equivalent for non-exceptional c (i.e., for h + c different from the integer values involved in Eq. (3.5)). For type I representations $\tilde{\ell} = \ell$ [= (0, ... ,0, ℓ) in the notation of Sec. 2A]. The intertwining mapping exhibiting this equivalence is given by (4.16) and (for type I representations by) (4.19). The (Fourier and) harmonic expansion of the intertwining kernel (4.19) is given by (5.23). For the exceptional points (with integer h+c) the representations (8.18) are, in general, only partially equivalent. There are additional intertwining differential operators in this case. The picture of intertwining mappings for the integer points is summarized by the quartet diagram of exact sequences of representation spaces' homomorphisms presented on Fig. 1 (Sec. 6B). The intertwining differential operators (defined in Sec. 6B) provide a link between invariant hermitian forms for subrepresentations and factor representations (Proposition 6.5 of Sec. 6 D).

APPENDIX A.

Symmetric tensor representations of SO(n) and their decomposition in IR's of SO(n -1)

A.1 Harmonic extension of homogeneous polynomial functions on the light cone

We have collected in this Appendix some facts about the type I representations of the orthogonal group, used in Sec. 5.B (for $n = 2h$). The results quoted in this first section are taken from [W1] [B2].

Proposition A.1. Every homogeneous polynomial.

$$f(z) = f^{\mu_1 \dots \mu_\ell} z_{\mu_1} \dots z_{\mu_\ell}, \quad z \in \mathbb{K}_n \tag{A.1}$$

on the cone (2.11) has a unique (homogeneous) harmonic extension $f_H(\zeta)$, $\zeta \in \mathbb{C}^n$ such that

$$\Delta_\zeta f_H(\zeta)[\equiv (\frac{\partial^2}{\partial \zeta_1^2} + \dots + \frac{\partial^2}{\partial \zeta_n^2}) f_H(\zeta)] = 0, \quad f_H(\zeta) = f_H^{\mu_1 \dots \mu_\ell} \zeta_{\mu_1} \dots \zeta_{\mu_\ell}, \tag{A.2}$$

$$f_H(z) = f(z) \text{ for } z \in \mathbb{K}_n. \tag{A.3}$$

The homogeneous polynomial $f_H(\zeta)$ is also determined by its values $f_H(\hat{\zeta})$ on the real unit sphere

$$S^{n-1} = \{\hat{\zeta} \in \mathbb{R}^n ; \hat{\zeta}^2 = 1\}, \tag{A.4}$$

which satisfy the equation

$$[\Delta_S + \ell(\ell + n - 2)] f_H(\hat{\zeta}) = 0, \tag{A.5}$$

where Δ_S is the Laplace-Beltrami operator on S^{n-1} [see (A.12) below]. The scalar product of two tensors $f_H^{\mu_1 \dots \mu_\ell}$ is proportional to the scalar product on $\mathcal{H}_{\ell}^{(n)} = \mathcal{L}^2(S^{n-1})$:

$$(f_{1H}, f_{2H})_{\mathcal{V}^{\ell}} \equiv \bar{f}_{1H\,\mu_1\ldots\mu_\ell} f_{2H}^{\mu_1\ldots\mu_\ell} =$$

$$= a_\ell \int_{S^{n-1}} \bar{f}_{1H}(\hat{\zeta}) f_{2H}(\hat{\zeta})\,(d\hat{\zeta}) \equiv a_\ell (f_1, f_2)_{\mathcal{H}_\ell^{(n)}} \tag{A.6}$$

where $(d\hat{\zeta})$ is the normalized surface element on the sphere.

Proof: The first statement is a consequence of the known fact that every homogeneous polynomial $P(\zeta)$ (of degree ℓ) can be expanded in a unique way as a sum of (homogeneous) harmonic polynomials $Y_s(\zeta)$ (see, e.g. p. 13 of [W1])

$$P(\zeta) = \sum_{0 \le 2k \le \ell} (\zeta^2)^k \, Y_{\ell-2k}(\zeta). \tag{A.7a}$$

To prove that we set

$$(\zeta^2)^k Y_{\ell-2k}(\zeta) = \prod_{\substack{j=0 \\ j\ne k}}^{[\frac{\ell}{2}]} \frac{\Omega - \lambda_{\ell-2j}}{\lambda_{\ell-2k} - \lambda_{\ell-2j}} P(\zeta) \tag{A.7b}$$

where

$$\Omega = \frac{1}{2}(\zeta_\mu \partial_\nu - \zeta_\nu \partial_\mu)(\zeta_\nu \partial_\mu - \zeta_\mu \partial_\nu) \tag{A.8}$$

is the Casimir operator of SO(n) with eigenvalues

$$\lambda_s (= \lambda_{ns}) = s(s+n-2) \tag{A.9}$$

corresponding to the eigenfunctions $(\zeta^2)^k Y_s(\zeta)$ (Ω commutes with ζ^2). We shall apply (A.7) to an arbitrary homogeneous extension $P(\zeta, f)$ of $f(\zeta)$ and shall identify $f_H(\zeta)$ with $Y_\ell(\zeta)$. To see that $Y_\ell(\zeta)$ is indeed harmonic, it suffices to note that the Laplacian $\Delta = \Delta_\zeta (= \sum_\mu \frac{\partial^2}{\partial \zeta_\mu^2})$ can be written in the form

$$\Delta = 4(\zeta^2)^{1-\frac{1}{2}n} \frac{\partial^2}{\partial \zeta^2}\left[(\zeta^2)^{\frac{n}{2}} \frac{\partial^2}{\partial \zeta^2}\right] - \frac{1}{\zeta^2}\Omega \tag{A.10}$$

and to use (A.9) with $s = \ell$. It remains to show that $Y_\ell(\zeta)$ only depends on $f(z)$ and not on the particular choice of its extension in $\mathbb{C}^n$. Let indeed $P_1(\zeta, f)$ and $P_2(\zeta, f)$ be two such extensions; then the difference $P_1(\zeta, f) - P_2(\zeta, f)$ vanishes on the cone $\zeta^2 (= z^2) = 0$, and hence has the form

$$P_1(\zeta, f) - P_2(\zeta, f) = \zeta^2 P_{\ell-2}(\zeta),$$

where $P_{\ell-2}(\zeta)$ is a homogeneous polynomial of degree $\ell-2$. The harmonic projection of this polynomial vanishes:

$$\zeta^2 \prod_{j=0}^{[\frac{\ell}{2}]-1} \frac{\Omega - \lambda_{\ell-2j}}{\lambda_\ell - \lambda_{\ell-2j}} P_{\ell-2}(\zeta) = 0.$$

The one to one relation between $f_H(\zeta)$ and its restriction to the real unit sphere (A. 4) is established trivially. Because of the homogeneity condition, f_H is extended to arbitrary real ζ by homogeneity

$$f_H(\zeta) = (\zeta^2)^{\ell/2} f_H(\hat{\zeta}) \text{ (for } \zeta^2 \neq 0\text{)}, \tag{A. 11}$$

and then to any complex ζ by analyticity. We note that Ω is related to the Laplace-Beltrami operator Δ_S on the sphere (A.4) by

$$\Omega = -\Delta_S. \tag{A. 12}$$

Finally, since the representation (ℓ) of SO(n), realized on the space $\mathcal{V}^\ell$ of symmetric traceless tensors, or equivalently, on the space $\mathcal{H}_\ell^{(n)}$ of harmonic polynomials on the unit sphere, is irreducible and since both scalar products $(\ ,\ )_{\mathcal{V}^\ell}$ and $(\ ,\ )_{\mathcal{H}_\ell^{(n)}}$ are invariant under the action of this representation given by (2. 12), Eq. (A. 6) follows from Schur's lemma. We shall evaluate the constant a_ℓ in Sec. A. 3, below.

Example: The harmonic extension of the polynomial $(b\mathfrak{z})^\ell$ $(\mathfrak{z}^2 = 0)$ is found from covariance consideration to be of the form $(b^2\zeta^2)^{\ell/2} P(\hat{b}\hat{\zeta})$ where $\hat{b}\hat{\zeta} \equiv (b^2\zeta^2)^{-\frac{1}{2}}(b\zeta)$ and P is a suitably normalized solution of (A. 5) regarded as a function of $\hat{\zeta}$. The result is:

$$H_{n\ell}(b,\zeta) = \frac{\ell!}{(\frac{n}{2}-1)_\ell} \left(\frac{1}{4} b^2 \zeta^2\right)^{\ell/2} C_\ell^{n/2-1}(\widetilde{b\zeta}) =$$

$$= (b\zeta)^\ell F\left(-\frac{\ell}{2}, \frac{1-\ell}{2} \; ; \; 2-\frac{n}{2}-\ell \; ; \; \frac{b^2\zeta^2}{(b\zeta)^2}\right) , \tag{A. 13}$$

where C_ℓ^ν and $F(\alpha, \beta; \gamma; x)$ are the Gegenbauer polynomial and the hypergeometric function, respectively, and the symbol $(a)_k$ is defined by (5. 11) .

A. 2 SO(n-1) expansion of homogeneous polynomials.

The zonal spherical functions

The existence and uniqueness of a decomposition of the type (A. 7) implies that each polynomial (of degree ℓ) on the unit sphere (A. 4) has a unique expansion in harmonic polynomials

$$P(\hat{\zeta}) = \sum_{s=0}^{\ell} Y_s(\hat{\zeta}) \qquad (\hat{\zeta} \in S^{n-1}) \tag{A. 14}$$

(where $Y_s(\hat{\zeta}) = Y_s^{(n)}(\hat{\zeta};P)$ satisfy (A. 9)) . Replacing n by $n-1$ (and S^{n-1} by S^{n-2}), we can use this result to obtain a SO(n-1) expansion for an arbitrary function $f(\mathfrak{z})$ of the type (A. 1). Indeed, giving to $\mathfrak{z}$ the special value:

$$\mathfrak{z} = (\underline{z}, i), \; \underline{z} = \bar{\underline{z}}, \; \underline{z}^2 = 1 \tag{A. 15}$$

(cf. (5. 9)), we express $f(\mathfrak{z})$ in terms of the (ℓ -th degree) polynomial $f(\underline{z}, i)$ on S^{n-2}, and by (A. 14) obtain

$$f(\underline{z}, i)\,[=(-i\mathfrak{z}_n)^{-\ell} f(\mathfrak{z})] = \sum_{s=0}^{\ell} Y_s^{(n-1)}(\underline{z}). \qquad (A.16)$$

Consider now the special case, in which $f(\mathfrak{z})$ is a covariant function of three vectors, say p, b and $\mathfrak{z}$ (and therefore depending only on their scalar products). By an appropriate choice of the coordinate system we can arrange that the vector p points along the n-th axis, while b lies in the $(n-1, n)$-plane. Then the function $f(\mathfrak{z})\ (= f(\mathfrak{z}\,; b, p))$ will be manifestly $SO(n-2)$ invariant [$SO(n-2)$ being the subgroup of $SO(n)$ acting non-trivially in the subspace $\mathbb{R}^{n-2}$ of $\mathbb{R}^n$ spanned by the first $(n-2)$ axes]. In this case, the harmonic polynomials $Y_{n-1}(\underline{z})$ in the expansion (A.16) depend on the single scalar variable

$$\omega = (b^2)^{-\frac{1}{2}}\ \underline{z}\,\underline{b}\quad (\equiv \cos \underline{z}\,\underline{b}) \qquad (A.17)$$

(cf. (5.4)). They are determined from the harmonicity property uniquely up to a constant factor. The normalized solution of

$$[\Omega[n-1] - s(s+n-3)]\,\hat{Y}_s^{(n-1)}(\omega) \equiv \left[(\omega^2-1)\frac{d^2}{d\omega^2} + (n-2)\omega\frac{d}{d\omega} - s(s+n-3)\right]\hat{Y}_s^{(n-1)}(\omega) = 0 \qquad (A.18)$$

given by

$$\hat{Y}_s^{(n-1)}(\omega) = (I_s^{(n-1)})^{-\frac{1}{2}}\, C_s^{\frac{n-3}{2}}(\omega) = (I_s^{(n-1)})^{-\frac{1}{2}}\,\frac{(n-3)_s}{(n/2-1)_s}\, P_s^{(n/2-2,\, n/2-2)}(\omega), \qquad (A.19)$$

where

$$I_s^{(n-1)} = \int_{S^{n-2}} [C_s^{\frac{n-3}{2}}(\underline{e}\,\underline{z})]^2 (d\underline{e}) = \frac{\sigma_{n-3}}{\sigma_{n-2}} \int_{-1}^{1} (1-\omega^2)^{\frac{n}{2}-2} [C_s^{\frac{n-3}{2}}(\omega)]^2 d\omega =$$

$$= \frac{\Gamma(\frac{n-1}{2})}{\Gamma(\frac{n}{2}-1)\pi^{1/2}}\ \frac{2^{4-n}\pi\,\Gamma(n-3+s)}{s!\,(s+\frac{n-3}{2})[\Gamma(\frac{n-3}{2})]^2} = \frac{1}{s!}\ \frac{n-3}{2s+n-3}\ \frac{(n+s-4)!}{(n-4)!} \qquad (A.20)$$

is called (normalized) zonal spherical function.

[In deriving (A. 20), we have used the formula

$$\sigma_\nu = \frac{2\pi^{\frac{\nu+1}{2}}}{\Gamma(\frac{\nu+1}{2})} \tag{A. 21}$$

for the surface of the unit sphere S^ν, Eq. 7.313.2 of ref. [G7] for the normalization integral of the Gegenbauer polynomial, and the doubling formula $2^{2\nu-1}\Gamma(\nu)\Gamma(\nu+1/2) = \pi^{1/2}\Gamma(2\nu)$ for Γ-functions (see Eq. 8.335.1 of ref. [G7]).]

A.3 Evaluation of the proportionality constant a_ℓ between the scalar products in $\mathcal{V}^\ell$ and in $\mathcal{H}_\ell^{(n)}$

In order to find the proportionality coefficient a_ℓ in (A.6), we shall evaluate the scalar products in both sides for the special case of vectors of the type (A.13). It follows from the definition of $H_{n\ell}(b,\zeta)$ that the usual square norm of the symmetric traceless tensor $(b^{\otimes \ell}-\mathrm{tr})$ is

$$(\bar{b}_{\mu_1}\cdots\bar{b}_{\mu_\ell}-\mathrm{tr})(b_{\mu_1}\cdots b_{\mu_\ell}-\mathrm{tr}) = H_{n\ell}(\bar{b}, b) \tag{A. 22}$$

Choosing for b a real unit vector e and using Eq. 8.937.4 of ref. [G7] we obtain

$$(H_{n\ell}(e,\zeta), H_{n\ell}(e,\zeta))_{\mathcal{V}^\ell} = H_{n\ell}(e,e) = \frac{\ell!\, C_\ell^{\frac{n}{2}-1}(1)}{(\frac{n}{2}-1)_\ell\, 2^\ell} =$$

$$= \frac{(n-2)_\ell}{2^\ell(\frac{n}{2}-1)_\ell} = 2^{-\ell}\frac{\Gamma(\frac{n}{2}-1)(n+\ell-3)!}{\Gamma(\frac{n}{2}+\ell-1)(n-3)!} \quad \left[(\alpha)_\ell = \frac{\Gamma(\alpha+\ell)}{\Gamma(\alpha)}\right]. \tag{A. 23}$$

On the other hand, according to (A. 20)

$$(H_{n\ell}(e,\zeta), H_{n\ell}(e,\zeta))_{\mathcal{H}_\ell^{(n)}} = \left[\frac{\ell!}{2^\ell (\frac{n}{2}-1)_\ell}\right]^2 I_\ell^{(n)} =$$

$$= \frac{\ell!}{2^{2\ell}[(\frac{n}{2}-1)_\ell]^2} \frac{n-2}{2\ell+n-2} \frac{(n+\ell-3)!}{(n-3)!} . \qquad (A.24)$$

From (A. 6), (A. 23) and (A. 24), we find

$$a_\ell = a_\ell^{(n)} = \frac{2^\ell (\frac{n}{2})_\ell}{\ell!} . \qquad (A.25)$$

A. 4. Derivation of a factorized expression for the projection operators $\sqcap^{\ell s}$

Here we shall use (for the first time) the canonical basis in $\mathcal{H}_\ell^{(n)}$ and shall write down some of the (generalized) spherical functions evaluated in Vilenkin [V1] .

Type I representations of SO(n-i) are labelled by just one number s_i ; therefore, we can denote the canonical basis vectors of the symmetric tensor representation (ℓ) of SO(n) by

$$\Xi^{(\ell)}_{s_1 \dots s_{n-2}} \equiv \Xi_S , \ell \geq s_1 \geq s_2 \dots \geq s_{n-3} \geq |s_{n-2}| . \qquad (A.26)$$

(Since the representation is fixed, we shall omit the superscript (ℓ) on the basis vectors.) The Ξ_S can be realized as harmonic functions on the unit sphere. To do that we define the rotation R_ζ in the plane

$(\hat{\zeta}, e_n)$ $(\hat{\zeta} \in S^{n-1}$, e_n is the unit vector of the nth axis) such that

$$R_\zeta e_n = \hat{\zeta} \quad , \qquad trR_\zeta = n-2+2(e_n, \hat{\zeta}) \equiv n-2+2\cos\varphi(\zeta) \ . \tag{A. 27}$$

Let

$$D^\ell_{SS'}(\Lambda) = \left\langle \Xi_S , D^\ell(\Lambda)\, \Xi_{S'} \right\rangle \tag{A. 28}$$

be the matrix elements of the representation D^ℓ in the canonical basis.

Then

$$\Xi^{(\ell)}_S(\hat{\zeta}) = d_n^{1/2}(\ell)\ D^\ell_{0S}(R_\zeta^{-1}) \tag{A. 29}$$

where $d_n(\ell)$ is the dimension of the representation (ℓ) given by

$$d_n^{(\ell)} = \binom{n+\ell-2}{\ell} \frac{n+2\ell-2}{n+\ell-2} \tag{A. 30}$$

(cf. (2. 2)). To prove that, we first note that the matrix elements $D^\ell_{0S}(\Lambda^{-1})$ only depend on the cosets

$$\Lambda\, SO(n-1) \in {}^{SO(n)}/SO(n-1) \ ,$$

since

$$D^\ell_{0S}(u^{-1}\Lambda^{-1}) = D^\ell_{0S}(\Lambda^{-1}) \quad \text{if} \quad u\, e_n = e_n \ (u \in SO(n-1) \subset SO(n)) \ ; \tag{A. 31}$$

this means that we can replace the rotation R_ζ in the definition (A. 29) by any Λ_ζ of the form $\Lambda_\zeta = R_\zeta\, u$, $u \in SO(n-1)_{e_n}$. Secondly, we observe that the rotation $R_{\Lambda^{-1}\zeta}$ belongs to the same coset as $\Lambda^{-1} R_\zeta$. Hence, the Ξ_S have the correct transformation law

$$D^{\ell}(\Lambda)\,\Xi_S(\hat{\zeta}) = d_n^{\frac{1}{2}}(\ell)\, D^{\ell}_{0S}(R_{\zeta}^{-1}\Lambda) = \sum_{S'} \Xi_{S'}(\hat{\zeta})\, D^{\ell}_{S'S}(\Lambda) \tag{A.32}$$

for basis vectors in the representation space $\mathcal{H}_{\ell}^{(n)}$. The normalization factor is chosen in such a way that the canonical basis vectors are (ortho) normalized:

$$\left\langle \Xi_S,\, \Xi_{S'} \right\rangle = \int \overline{\Xi}_S(\zeta)\, \Xi_{S'}(\zeta)\, (d\zeta) = \delta_{SS'}\,,$$
$$\left(\int_{S^{n-1}} (d\hat{\zeta}) = 1 \right). \tag{A.33}$$

(This follows from known normalization properties of matrix elements of representations of compact group, -- see, e.g., [30], Sec. 27.)

Each vector $f = f(\hat{\zeta}) \in \mathcal{H}_{\ell}^{(n)}$ can be written in the form

$$f(\hat{\zeta}) = \sum_{S} \Xi_S(\hat{\zeta}) \left\langle \Xi_S,\, f \right\rangle = d_n^{\frac{1}{2}}(\ell) \left\langle D^{\ell}(R_{\zeta})\, \Xi_0,\, f \right\rangle. \tag{A.34}$$

In deriving (A.34) we have used the unitarity of the representation $D^{\ell}(\Lambda)$, which implies the identities:

$$\overline{D^{\ell}_{0S}(R^{-1}_{\zeta'})} = D^{\ell}_{S0}(R_{\zeta'})\,, \qquad \overline{\Xi_0(R^{-1}_{\zeta'} R_{\zeta})} = \Xi_0(R^{-1}_{\zeta} R_{\zeta'})$$

Let A be any operator in $\mathcal{H}_{\ell}^{(n)}$ with kernel $\mathcal{A}(\hat{\zeta}, \hat{\zeta}')$:

$$(Af)(\hat{\zeta}) = \int \mathcal{A}(\hat{\zeta}, \hat{\zeta}')\, f(\hat{\zeta}')\, (d\hat{\zeta}'). \tag{A.35}$$

Then the kernel $\mathcal{A}$ is given by

$$\mathcal{A}(\hat{\zeta}, \hat{\zeta}') = d_n(\ell) \left\langle \Xi_0,\, D^{\ell}(R^{-1}_{\zeta})\, A D^{\ell}(R_{\zeta'})\, \Xi_0 \right\rangle. \tag{A.36}$$

Indeed, noting the connection between integration on the unit sphere S^{n-1} and invariant integration on the group SO(n) and using (A. 34), (A. 32) and the orthogonality relations

$$\int D^{\ell}_{s_1 s_2}(\Lambda)\, \overline{D^{\ell}_{s_3 s_4}(\Lambda)}\, d\Lambda = d_n^{-1}(\ell)\, \delta_{s_1 s_3}\, \delta_{s_2 s_4}, \tag{A. 37}$$

([22], Sec. 27, p. 73), we derive

$$\int \mathcal{A}(\hat{\zeta}, \hat{\zeta}')\, f(\hat{\zeta}')\, (d\hat{\zeta}') = d_n^{3/2}(\ell) \int \langle A^* D^{\ell}(R_{\zeta})\, \Xi_0 , D^{\ell}(\Lambda)\, \Xi_0 \rangle \langle D^{\ell}(\Lambda)\Xi_0, f \rangle$$

$$= d_n^{1/2}(\ell) \langle D^{\ell}(R_{\zeta})\, \Xi_0 , Af \rangle = (Af)(\zeta) .$$

In order to apply (A. 36) to the projection operator $\sqcap^{\ell s} = \sqcap^{\ell s}(e_n)$ (which projects onto the invariant subspaces $\mathcal{H}_s^{(n-1)}$ of $SO(n-1)_{e_n}$) we shall first write down the rotation R_{ζ} (A. 27) in the form

$$R_{\zeta} = u_{\zeta}\, R(\varphi)\, u_{\zeta}^{-1} , \tag{A. 38a}$$

where

$$R(\varphi) = \begin{pmatrix} \mathbf{1}_{n-2} & 0 & 0 \\ 0 & \cos\varphi & \sin\varphi \\ 0 & -\sin\varphi & \cos\varphi \end{pmatrix}, \cos\varphi = (e_n, \hat{\zeta}) , \quad u_{\zeta} \in SO(n-1). \tag{A. 38b}$$

Then $\Lambda_{\zeta} = U_{\zeta} R(\varphi)$ belongs to the same coset as R_{ζ} and the kernel $\mathcal{T}^{\ell s}$ of the projection operator can be written in the following manifestly positive form:

$$\Pi^{\ell s}(\hat{\zeta}, \hat{\zeta}') = d_n(\ell) \left\langle \Xi_0, D^{\ell}(R^{-1}(\varphi)) \sqcap^{\ell s} D^{\ell}(u_{\zeta}^{-1} u_{\zeta'}) D^{\ell}(R(\varphi')\Xi_0 \right\rangle =$$

$$= d_n(\ell) D^{\ell(n)}_{0s}(R^{-1}(\varphi)) D^{s(n-1)}_{00}(u_{\zeta}^{-1} u_{\zeta'}) D^{\ell(n)}_{s0}(R(\varphi')) \qquad (A.39)$$

Here $D^{\ell(k)}$ $(k = n, n-1)$ stands for the representation matrix of the IR (ℓ) of $SO(k)$, and our notation does not distinguish between an element u of $SO(n-1)$ and its canonical imbedding in $SO(n)$.

In order to evaluate explicitly the right-hand side of (A. 39) we set

$$\hat{\zeta} = (z \sin\varphi, \cos\varphi) \qquad \hat{\zeta}' = (z' \sin\varphi', \cos\varphi') \qquad (A.40a)$$

$$z^2 = z'^2 = 1, \qquad z z' = \omega (= \cos\theta). \qquad (A.40b)$$

The matrix elements appearing in (A. 39) are evaluated in [V1]. In particular,

$$D^{s(n-1)}_{00}(u_{\zeta}^{-1} u_{\zeta'}) = \frac{s!}{(n-3)_s} C_s^{n/2-3/2}(\omega), \qquad (A.41)$$

is proportional to the zonal spherical function (A. 19) (it is denoted by $P^{n-1\,s}_{000}(\omega)$ in Vilenkin [V1]):

$$D^{\ell(n)}_{0s}(R^{-1}(\varphi)) = D^{\ell(n)}_{s0}(R(\varphi)) =$$

$$= 2^s \left(\frac{n}{2}-1\right)_s \left[\frac{\ell!(\ell-s)!}{s!} \frac{2s+n-3}{(s+n-3)_{\ell+1}(n-2)_{\ell}}\right]^{1/2} \sin^s\varphi\, C^{n/2+s-1}_{\ell-s}(\cos\varphi). \qquad (A.42)$$

Combining together (A. 39) - (A. 42) (A. 30) and taking into account that

$$\Pi^{\ell s}(\hat{\zeta}, \hat{\zeta}) = \frac{2^{\ell}\left(\frac{n}{2}\right)_{\ell}}{\ell!} \hat{\zeta}^{\mu_1} \ldots \hat{\zeta}^{\mu_{\ell}} \sqcap^{\ell s}_{\mu_1 \ldots \mu_{\ell}\, \nu_1 \ldots \nu_{\ell}} \hat{\zeta}'^{\nu_1} \ldots \hat{\zeta}'^{\nu_{\ell}} \qquad (A.43)$$

(because of (A. 6) (A. 25)) and identifying $\cos\varphi$ with $\hat{p}\zeta_1$, $\cos\varphi'$ with $\hat{p}\zeta_2$ we end up with Eq. (5. 21).

A.5 Interior differentiation on the complex cone. Expression for the convolution of two tensors in terms of homogeneous polynomials

There are two equivalent ways to write down the convolution product of a rank ν tensor f with a rank ℓ tensor g ($\nu \leq \ell$) in the language of homogeneous polynomials. One (used throughout Sec. 6) consists in taking the harmonic extension $g(\zeta)$ of $g(\mathfrak{z})$ and then noting that

$$(f*g)(\zeta) = f_{\mu_1\dots\mu_\nu}\, g_{\mu_1\dots\mu_\nu\mu_{\nu+1}\dots\mu_\ell}\,\zeta_{\mu_{\nu+1}}\cdots\zeta_{\mu_\ell} = \frac{(\ell-\nu)!}{\ell!}\, f(\partial_\zeta)\,g(\zeta) \tag{A.44}$$

We shall describe in what follows another technique which does not need a harmonic extension. It uses instead an interior differentiation D_μ on the cone (2.11) (see [B2]). D_μ is defined up to a factor by demanding that it is the lowest order differential operator with the following three properties: (i) it is a lowering operator, – that is, it maps the space $\mathcal{V}^\ell$ of homogeneous polynomials of the type (2.10) for $\ell = 1,2,\dots$ into $\mathcal{V}^{\ell-1}$; (ii) it is an n-vector, in particular it has the same commutation relations as $\mathfrak{z}_\mu$ with the generators $X_{\mu\nu}$ of rotation (i.e. $[X_{\mu\nu}, \mathcal{D}_\lambda] = \delta_{\mu\lambda}\mathcal{D}_\nu - \delta_{\nu\lambda}\mathcal{D}_\mu$ -cf. (1.6'); (iii) it is an interior operator on the cone (2.11), – that is, for any polynomial $f(\mathfrak{z})$ on $\mathbb{K}_n$ we have

$$\mathcal{D}_\mu\,[\mathfrak{z}^2 f(\mathfrak{z})] = 0 \qquad \text{for} \quad \mathfrak{z}^2 = 0 \tag{A.45}$$

Examples of interior operators on $\mathbb{K}_n$ (which are, however, not lowering operators) are given by the generators of rotation $X_{\mu\nu} = s_{\mu\nu}$ (3.31) and dilatations

$$X = h - 1 + \mathfrak{z}\partial \qquad \left(\partial \equiv \frac{\partial}{\partial\mathfrak{z}}\ ,\ h = \tfrac{1}{2}n\right) \tag{A.46}$$

on the complex cone. There is no first order lowering interior differentiation, since $\partial_\mu \mathfrak{z}^2 = 2\mathfrak{z}_\mu$ ($\neq 0$ for $\mathfrak{z}^2 = 0$). We shall show that there is an unique up to a constant factor second order operator $\mathcal{D}_\mu$, satisfying (i) – (iii). We first study the uniqueness claim. The most general second order operator satisfying (i) and (ii) is

$$\mathcal{D}_\mu = a\,\partial_\mu + b\,X\partial_\mu + c\,\mathfrak{z}_\mu\,\Delta \qquad , \ (\Delta = \partial_1^2 + \dots + \partial_n^2)\,.$$

Applying (A.45) with f = 1 and f = $\mathfrak{z}_\nu$, we find

$$(a + bh + 2ch)\, 2\mathfrak{z}_\mu = 0$$

$$[a + b(h+1) + 2(h+1)c]\, 2\mathfrak{z}_\mu \mathfrak{z}_\nu = 0$$

hence, $c = -\frac{1}{2}b$, $a = 0$, so that

$$D_\mu = b\,(X\partial_\mu - \tfrac{1}{2}\mathfrak{z}_\mu \Delta) \tag{A.47}$$

The operator so defind is indeed interior, since for every polynomial $f(\mathfrak{z})$ one has the relation

$$D_\mu [\mathfrak{z}^2 f(\mathfrak{z})] = \mathfrak{z}^2 (D_\mu + 2b\partial_\mu) f(\mathfrak{z}) \tag{A.48}$$

so that (A.45) is verified.

In what follows we shall choose the normalization constant

$$b = 1 \tag{A.49}$$

With this choice the operators $X_{\mu\nu}$ ·

$$X_{0\mu} = \frac{1}{\sqrt{2}}(\mathfrak{z}_\mu - D_\mu)\,,\quad X_{\mu,n+1} = \frac{1}{i\sqrt{2}}(\mathfrak{z}_\mu + D_\mu) \tag{A.50}$$

and $X_{0,n+1} = -iX$ are the (mathematical) generators of the (real) Lie algebra of the conformal group $SO_0(n,2)$. That follows from the easily verifyable commutation relations

$$[D_\mu, \mathfrak{z}_\nu] = X\delta_{\mu\nu} + X_{\mu\nu} \tag{A.51}$$

With the operators (A.47) (A.49) we can replace Eq. (A.44) for the covolution product by

$$f(D)\,g(\mathfrak{z}) = (h + \ell - \nu - 1)_\nu \frac{\ell!}{(\ell-\nu)!}(f * g)(\mathfrak{z})\,. \tag{A.52}$$

We note that the vector operator $D = (D_1, \ldots D_{2h})$ satisfies the identity

$$D^2 \;(= D_1^2 + \ldots + D_{2h}^2) = 0 \tag{A.53}$$

hence , we do not need the harmonic extension of either factor in the left -hand side of (A.52).

It is useful to have an explicit formula for the action of $(pD)^\nu$ on a homogeneous polynomial in $\mathfrak{z}$ of degree $k \geqslant \nu$ (where p is a fixed 2h-vector). We note that

$$(pD)^\nu = \prod_{i=1}^{\nu} [(h + k - 1 - i)(p\partial) - \tfrac{1}{2} p\mathfrak{z}\Delta] \tag{A.54}$$

Obviously (A.54) is a polynomial in p^2, $p\mathfrak{z}$, $p\partial$ and Δ. Accounting for the degrees of homogeneity in p and $\mathfrak{z}$ we can write the following general expression for (A.54):

$$(pD)^{\nu} = \sum_{m,n} a(\nu, m, n)\, (\tfrac{1}{2}p^2)^n (p\mathfrak{z})^{m-n} (p\partial)^{\nu-m-n} (-\tfrac{1}{2}\Delta)^m \tag{A.55}$$

where the sum runs over nonnegative exponents ($0 \leq n \leq m \leq \nu - n$) and the coefficients $a(\nu, m, n)$ may also depend on k and h. Applying to both sides of (A.55) the operator pD we find the following recurrence relation for the a's:

$$\begin{aligned} a(\nu+1, m, n) &= (m-n+1)(2h+2k-2\nu-4-m+n)\, a(\nu, m, n-1) + \\ &+ (h+k-\nu-2-m+n)\, a(\nu, m, n) + a(\nu, m-1, n) \qquad \begin{array}{l} 1 \leq m \leq \nu \\ 1 \leq n \leq \min(\nu-m, m-1) \end{array} \end{aligned} \tag{A.56}$$

which have to be supplemented by the boundary conditions

$$\begin{aligned} a(\nu+1, m, 0) &= (h+k-\nu-2-m)\, a(\nu, m, 0) + a(\nu, m-1, 0), \quad 1 \leq m \leq \nu \\ a(\nu+1, m, m) &= (h+k-\nu-2)\left[2a(\nu, m, m-1) + a(\nu, m, m)\right], \quad 1 \leq m \leq \tfrac{\nu}{2} \\ a(\nu+1, m, \nu+1-m) &= (2m-\nu)(2h+2k-\nu-2m-3)\, a(\nu, m, \nu-m) + a(\nu, m-1, \nu+1-m), \quad \tfrac{\nu}{2}+1 \leq m \leq \nu \\ a\left(\nu+1, \tfrac{\nu+1}{2}, \tfrac{\nu+1}{2}\right) &= 2(h+k-\nu-2)\, a\left(\nu, \tfrac{\nu+1}{2}, \tfrac{\nu-1}{2}\right), \quad \nu \text{ odd} \\ a(\nu, 0, 0) &= (h+k-\nu-1)_{\nu} \\ a(\nu, \nu, 0) &= 1 \end{aligned} \tag{A.57}$$

Since it is difficult to solve directly the resulting system, we shall apply both sides of (A.55) to the simple homogeneous polynomial $(b\mathfrak{z})^k$ where b is some fixed 2h-vector. The right-hand side gives

$$(pD)^{\nu}(b\mathfrak{z})^k = k! \sum_{m,n} \frac{a(\nu, m, n)}{(k-\nu-m+n)!} \left(\frac{p^2}{2}\right)^n (p\mathfrak{z})^{m-n} (pb)^{\nu-m-n} \left(-\frac{b^2}{2}\right)^m (b\mathfrak{z})^{k-\nu-m+n} \tag{A.58}$$

To evalutate the left-hand side we use (A.13):

$$\begin{aligned} (pD)^{\nu}(b\mathfrak{z})^k &= \left[(pD)^{\nu} H_{2h,k}(b, \mathfrak{z})\right]_{\mathfrak{z}=\mathfrak{z}} = \frac{\Gamma(h+k-1)}{\Gamma(h+k-\nu-1)} \left[(p\partial)^{\nu} H_{2h,k}(b, \mathfrak{z})\right]_{\substack{\mathfrak{z}=\mathfrak{z} \\ \mathfrak{z}^2=0}} = \\ &= \frac{k!\,\nu!}{\Gamma(h+k-\nu-1)} \sum_{m,n} \frac{\Gamma(h+k-1-m) \left(\frac{p^2}{2}\right)^n \left(-\frac{b^2}{2}\right)^m (pb)^{\nu-m-n} (p\mathfrak{z})^{m-n} (b\mathfrak{z})^{k-\nu-m+n}}{(\nu-n-m)!\,(k-\nu-m+n)!\,(m-n)!\,n!} \end{aligned} \tag{A.59}$$

Comparing (A.58) and (A.59) we obtain:

$$a(\nu,m,n) = \frac{\nu!\,(h+k-\nu-1)_{\nu-m}}{(\nu-m-n)!\,(m-n)!\,n!} \tag{A.60}$$

It is easy to verify that the coefficients (A.60) satisfy the recurrence relations (A.56) and (A.57).

APPENDIX B.

The special cases h = 1 and h = $\frac{1}{2}$. Relation to the formalism of two by two matrices

B.1 Splitting of the representations χ of $O^{\uparrow}(3,1)$ into elementary representations of SL(2, C)

It turns out that the elementary representations $[\ell, c]$ of $O^{\uparrow}(3,1)$ are reducible for $\ell > 0$, when restricted to the proper Lorentz group $SO^{\uparrow}(3,1)$. The universal covering Spin (3,1) of $SO^{\uparrow}(3,1)$ is isomorphic to the group SL(2, C) of 2x2 complex unimodular matrices studied thoroughly by Gel'fand, et al. [G2].

In order to exhibit the precise relation between the representations of $O^{\uparrow}(3,1)$ and of SL(2, C), we shall first recall the definition (and labelling) of the elementary representations of SL(2, C) adopted in ref. [G2].

To each pair of complex numbers (n_1, n_2), whose difference $n_1 - n_2$ is an integer, we make correspond a respresentation of SL(2, C), acting in an appropriate space $D_{(n_1, n_2)}$ of infinitely smooth functions $\varphi(z)$ of a complex variable $z = x_1 + ix_2$:

if $g = \begin{pmatrix} \alpha & \beta \\ \gamma & \delta \end{pmatrix} / \alpha\delta - \beta\gamma = 1 /$, then $g^{-1} = \begin{pmatrix} \delta & -\beta \\ -\gamma & \alpha \end{pmatrix}$

$$\left[T_{n_1, n_2}(g)\varphi\right](z) = (\alpha - \gamma z)^{n_1 - 1} (\overline{\alpha - \gamma z})^{n_2 - 1} \varphi\left(\frac{\delta z - \beta}{\alpha - \gamma z}\right). \qquad \text{(B.1)}$$

(We have converted the right translations used in [G2], into the left translations used throughout this paper.)

We shall show that each of the representations $\chi = [\ell > 0, c]$ of $\overset{\uparrow}{O}(3,1)$ (see Sec. 2D) can be decomposed into two representations (n_1, n_2) of the type (B.1):

$$[\ell, c] = (n_1, n_2) \oplus (n_2, n_1) \quad \text{where} \quad n_1 - n_2 = 2\ell, \; n_1 + n_2 = -2c. \tag{B.2}$$

In order to prove this statement, we first notice that each symmetric traceless tensor $f_{\mu_1 \ldots \mu_\ell}$ ($\ell = 1, 2, \ldots$) in two dimensions ($\mu_i = 1, 2$) has just two independent components, say, $f_{1\ldots 11}$ and $f_{1\ldots 12}$. For each $f_{\mu_1 \ldots \mu_\ell}(x_1, x_2) \in C_\chi$ we define a pair of functions $\varphi_+(z) \in D_{(n_1, n_2)}$ and $\varphi_-(z) \in D_{(n_2, n_1)}$ by

$$\varphi_\pm(x_1 \pm i x_2) = f_{1\ldots 11}(x_1, x_2) \pm i f_{1\ldots 12}(x_1, x_2). \tag{B.3}$$

Then, it is straightforward to verity that the transformations (2.31 a, b) for f go into appropriate transformations of type (B.1) for the φ's. For instance, dilatations correspond to matrices

$$a = \begin{pmatrix} |a|^{1/2} & 0 \\ 0 & |a|^{-1/2} \end{pmatrix}, \tag{B.4a}$$

so that

$$\left[T_{n_1 n_2}(a) \varphi_\pm \right](z) = |a|^{-1-c} \varphi_\pm(z/|a|), / c = -\tfrac{1}{2}(n_1 + n_2) /, \tag{B.4b}$$

---

* The notation (n_1, n_2) used in this Appendix should not be confused with the symbol for the highest weight of an IR of SO(n) used in Sec. 2A.

while rotations (on angle θ) in the (1, 2) -plane are represented by

$$\left[T_{n_1 n_2}(g_\theta)\,\varphi_\pm\right](z) = e^{\pm i\ell\theta}\,\varphi_\pm(e^{-i\theta} z)\,, \tag{B.5a}$$

where

$$g_\theta = \begin{pmatrix} e^{i\theta/2} & 0 \\ 0 & e^{-i\theta/2} \end{pmatrix}, \quad \ell = \frac{1}{2}(n_1 - n_2)\,. \tag{B.5b}$$

On the other hand the conformal inversion (2.31c) goes into $Rz = -1/\bar{z}$ which is not an SL(2, C) transformation. It could be replaced by the proper rotation RI_2, where $I_2(x_1, x_2) = (x_1, -x_2)$; we have

$$g_{RI_2} = \begin{pmatrix} 0 & 1 \\ -1 & 0 \end{pmatrix}; \quad T_{(n_1, n_2)}(g_{RI_2})\,\varphi_\pm(z) = \frac{1}{(z\bar{z})^{1+c}}\,\varphi_\pm\left(-\frac{1}{z}\right). \tag{B.6}$$

The operator of space reflection $T_\chi(I_2)$ in C_χ mixes the two terms in the direct sum decomposition (B. 2):

$$\left[T_\chi(I_2)\,\varphi_\pm\right](z) = \varphi_\mp(I_2 z)\,, \quad I_2 z = \bar{z}\,. \tag{B.7}$$

For $\ell = 0$ the representation $[0, c]$ is equivalent to the representation (c, c) of SL(2, C).

B. 2 Vanishing of the projection operators $\prod^{\ell s}$ for $s > 1$

We already observed that for $h = 1$ and any $\ell > 0$ the number of independent components of $f_{\mu_1 \ldots \mu_\ell}$ is two. Therefore, we can have at most two non-vanishing projection operators in the momentum space expansion (5.23) of the invariant two-point function. Actually, we shall show, that for $h = 1, \ell > 0$ the operators $\Pi^{\ell 0}$ and $\Pi^{\ell 1}$ are 1-dimensional projectors, while all other $\Pi^{\ell s}$ (with $s > 1$) vanish.

To see that, we notice that the variable ω in (5.4) or (5.21c) is equal to -1 for $h = 1$. Indeed, since ω in (5.4) is a homogeneous function of $\mathfrak{z}_1$ and $\mathfrak{z}_2$ we can take (for non-collinear $\mathfrak{z}$'s) $\mathfrak{z}_1 = (1, i)$, $\mathfrak{z}_2 = (1, -i)$ so that

$$\omega = 1 - \frac{2p^2}{|p_1 + ip_2|^2} = -1 \quad . \tag{B.8}$$

Furthermore, we notice that

$$C_s^{-1/2}(-1) = (-1)^s \, C_s^{-1/2}(1) = 0 \qquad \text{for} \quad s > 1 ; \tag{B.9}$$

therefore, according to (5.21c) $\Pi^{ss}(p) = 0$ for $s > 1$. Inserting in (5.21a) we find

$$\begin{aligned}
\Pi^{00}(p) &= 1 , \\
\Pi^{\ell 0}(p) &= 2^{\ell-1} L_{\ell 0}(p) \, L_{\ell 0}(p) , \quad \ell \geqslant 1 , \\
\Pi^{\ell 1}(p) &= 2^{\ell-1} L_{\ell 1}(p) \quad L_{\ell 1}(p) \, \Pi^{11}(p) , \\
\Pi^{\ell s}(p) &= 0 \quad \text{for} \quad s > 1 .
\end{aligned} \tag{B.10}$$

B. 3 The structure of exceptional representations for h = 1

First of all, we observe that the structure of the representations $\chi^{\pm}_{\ell\nu}$ remains qualitatively the same for $h=1$. However, the situation changes for the pair $\chi'^{+}_{\ell\nu}$. We shall show that for $\ell \geq 1$ the representations $\chi'^{\pm}_{\ell\nu}$ are irreducible.

To see that, we notice that

$$C'^{-}_{\ell\nu} = F'_{\ell\nu} \quad \text{for} \quad h = 1 , \; \ell > 0 . \tag{B. 11}$$

Indeed, the (sub) space $F'_{\ell\nu}$ can be defined as the kernels of the intertwining operator $G'^{+}_{\ell\nu}$ (6. 13). On the other hand, it follows from (6. 13) and from Section B. 2. that

$$G'^{+}_{\ell\nu} \equiv 0 \quad \text{for} \; h=1 , \; \ell \geq 1 \; (\nu = 1, 2, \ldots) . \tag{B. 12}$$

This proves (B. 11). It also shows that the subspace $D_{\ell\nu} \subset C'^{+}_{\ell\nu}$ is trivial:

$$D_{\ell\nu} = \{0\} \quad \text{for} \; h=1 , \; \ell \geq 1 , \tag{B. 13}$$

since $D_{\ell\nu}$ is the image of $C'^{-}_{\ell\nu}$ under the mapping $G'^{+}_{\ell\nu}$.

On the other hand, the representations $\chi'^{\pm}_{0\nu}$ are reducible (like in the general case) $F'_{\ell\nu}$ ($D_{\ell\nu}$) being nontrivial invariant subspace of $C'^{-}_{\ell\nu}(C'^{+}_{\ell\nu})$. This is not true, however, for the SL(2, C) representations $(\nu, -\nu)$ and $(-\nu, \nu)$ appearing in the direct sum decomposion

$$\chi'^{+}_{0\nu} = \chi'^{-}_{0\nu} = [\, \nu , 0 \,] = (\nu, -\nu) \oplus (-\nu, \nu) . \tag{B. 14}$$

To see what happens we consider the special case $\nu = 1$. In this case, the invariant subspace $D_{01} \subset C'^{+}_{01}$ is defined as the set of vector functions $f_{\mu}(x) \in C'^{+}_{01}$ which satisfy the transversality condition

$$\partial_{\mu} f_{\mu}(x) = 0 . \qquad \text{(B. 15)}$$

Using (B. 3), we see that (B. 15) is equivalent to

$$\frac{\partial}{\partial z} \varphi_{+}(z) + \frac{\partial}{\partial \bar{z}} \varphi_{-}(z) = 0 \qquad \text{(B. 16)}$$

and, thus, mixes the two representations (1, -1) and (-1, 1).

For $h = 1$, $\ell > 0$ the first quartet diagram on Fig. 1 becomes degenerate and can be replaced by the following simpler diagram of exact (directed) sequences:

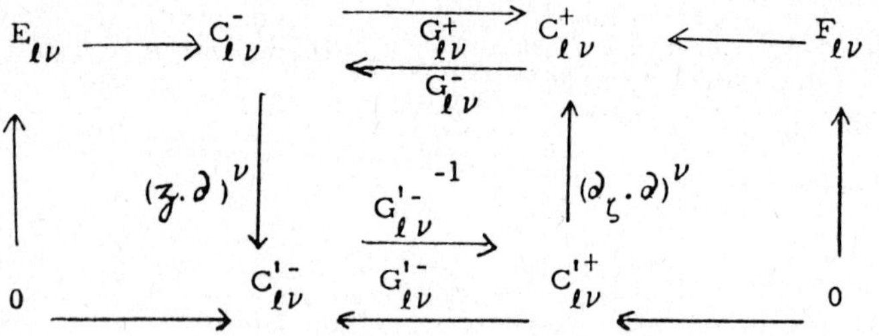

Fig. 3

The quartet diagram for the Lorentz group ($h = 1$) and $\ell > 0$

Finally we note that although the diagram on Fig. 3 has a similar structure as the diagram on Fig. 4 (Sec. 3. 3 of Chap. III) of ref. [G2], it has a different content, since each representation $\chi^{(')\pm}_{\ell\nu}$ is again a sum of two representations (n_1, n_2) of SL(2, C), and to each arrow in our picture corresponds a pair of arrows in the picture of Gel'fand, et al. For instance, the homomorphism $(z \cdot \partial)^{\nu} : C^{-}_{\ell\nu} \to C'^{-}_{\ell\nu}$ is split as follows in SL(2, C) variables:

$$C^{-}_{\ell\nu} = \begin{array}{c} D_{(2\ell+\nu,\,\nu)} \xrightarrow{(\partial/\partial z)^{2\ell+\nu}} D(-2\ell-\nu,\,\nu) \\ \oplus \qquad\qquad\qquad \oplus \\ D_{(\nu,\,2\ell+\nu)} \xrightarrow{(\partial/\partial \bar{z})^{2\ell+\nu}} D_{(\nu,\,-2\ell-\nu)} \end{array} = C'^{-}_{\ell\nu} \;. \qquad \text{(B.17)}$$

B. 4 Elementary representations of $SO^{\uparrow}(2, 1)$. The analytic discrete series

The elementary representations of $SO^{\uparrow}(2,1)$ are labelled by a single number c, since M is trivial in this case. They correspond to the representation $[\,\varepsilon = 0\,,\ s = -\,2c]$ of the two-fold covering group $SL(2,\mathbb{R})$ in the notation of ref. [G2]. (The representations of $SL(2,\mathbb{R})$ with $\varepsilon = 1$ also studied in [B1,G2] are double valued representation of $SO^{\uparrow}(2, 1)$.)

If we set again $g = \begin{pmatrix}\alpha & \beta\\ \gamma & \delta\end{pmatrix} \in SL(2,\mathbb{R})$ ($\alpha, \beta, \gamma, \delta$ - real) then the elementary representation T is given by its action on functions $f(x)$ of a single real variable x according to

$$[\,T^{c}(g)\,f]\,(x) = (\alpha - \gamma\, x)^{-1\,-2c}\ f\,\Big(\frac{\delta\, x - \beta}{\alpha - \gamma\, x}\Big)\,. \qquad \text{(B.18)}$$

They are irreducible except for the half integer points

$$\pm\, c = \ell + h - 1 = \ell - \tfrac{1}{2} \qquad \ell = 1,\ 2,\ \ldots \qquad \text{(B.19)}$$

The negative type representations (with $c = \frac{1}{2} - \ell$) contain the finite dimensional invariant subspaces E_{ℓ} of the polynomials of degree $2\ell - 2$. The vectors of E_{ℓ} are annihilated by the intertwining differential operator $\nabla_{x}^{2\ell\,-1}$, which maps the space $C_{\frac{1}{2}-\ell}$ onto the invariant subspace D_{ℓ} of $C_{\ell-\frac{1}{2}}$ of functions $f(x)$, satisfying

$$\int_{-\infty}^{\infty} f(x)\, x^k\, dx = 0 \qquad k = 0, \ldots, 2\ell - 2 . \tag{B. 20}$$

The representation $T^{\ell - \frac{1}{2}}$ of $O^{\uparrow}(2,1)$ is irreducible on D_ℓ. However, it splits into two irreducible subrepresentations T_ℓ^+ and T_ℓ^- when restricted to the identity component $G = SO^{\uparrow}(2, 1)$ of the three dimensional Lorentz group. The space D_ℓ then splits into two invariant subspaces D_ℓ^+ and D_ℓ^- which consists of boundary values of analytic functions in the upper and lower half planes $z = x + iy$, respectively. The IR's $T_\ell^{\pm}$ on $D_\ell^{\pm}$ form the (analytic) discrete series of unitary representations of G. Since they are analogous to each other we shall restrict ourselves to a brief discussion of T_ℓ^+.

In order to make more transparent the interrelation between the K-induced and NAM-induced pictures of the discrete series, considered in Sec. 7, we shall outline the corresponding results in the simple case at hand.

It is convenient to use the $SL(2, \mathbb{R})$ notation as in (B.18); then G can be defined as the factor group $SL(2, \mathbb{R})/\mathbb{Z}_2$.

Each factor in the Iwasawa decomposition (1.12) is a one-parameter subgroup of $\tilde{G} = SL(2, \mathbb{R})$

$$\tilde{n}_x = \begin{pmatrix} 1 & x \\ 0 & 1 \end{pmatrix}, \quad a_y = \begin{pmatrix} \sqrt{y} & 0 \\ 0 & \frac{1}{\sqrt{y}} \end{pmatrix}, \quad k_\theta = \begin{pmatrix} \cos\frac{\theta}{2} & \sin\frac{\theta}{2} \\ -\sin\frac{\theta}{2} & \cos\frac{\theta}{2} \end{pmatrix}; \tag{B. 21}$$

the decomposition (1.12) assumes the form

$$g = \tilde{n}_x a_y k_\theta = \frac{1}{\sqrt{y}} \begin{pmatrix} y & x \\ 0 & 1 \end{pmatrix} \begin{pmatrix} \cos\frac{\theta}{2} & \sin\frac{\theta}{2} \\ -\sin\frac{\theta}{2} & \cos\frac{\theta}{2} \end{pmatrix} =$$

$$= y^{-\frac{1}{2}} \begin{pmatrix} y\cos\frac{\theta}{2} - x\sin\frac{\theta}{2} & y\sin\frac{\theta}{2} + x\cos\frac{\theta}{2} \\ -\sin\frac{\theta}{2} & \cos\frac{\theta}{2} \end{pmatrix} \tag{B. 22}$$

We consider the space $\mathcal{L}^2_\ell$ of square integrable functions on G, which satisfy the covariance condition

$$\mathfrak{f}(g k_\theta) = e^{i\ell\theta}\, \mathfrak{f}(g) . \tag{B. 23}$$

These functions are defined by their values on the homogeneous space $G/K \approx \tilde{N}A$, which is isomorphic to the upper half-plane $C_+ = \{z = x+iy\ ;\ y > 0\}$. The representation U^+_ℓ acts on such function as a left (quasi) regular representation [see (7. 7)]. Introducing the functions $\phi(\vec{x})$ (7. 12) on the homogeneous space C_+ :

$$\phi(\vec{x}) \equiv \phi(z) = \mathfrak{f}(\tilde{n}_x a_y), \quad z = x + iy \tag{B. 24}$$

one derives the following transformation law for the ϕ's :

$$[U^+_\ell(g)\,\phi](z) = \left(\frac{|\alpha - \gamma z|}{\alpha - \gamma z}\right)^{2\ell} \phi\left(\frac{\delta z - \beta}{\alpha - \gamma z}\right) \quad \text{for} \quad g = \begin{pmatrix} \alpha & \beta \\ \gamma & \delta \end{pmatrix} . \tag{B. 25}$$

Eq. (B. 24) maps $\mathcal{L}^2_\ell$ onto the Hilbert space $\mathcal{H}_\ell$ of functions ϕ with invariant scalar product

$$(\phi_1 , \phi_2) = \int \overline{\phi_1 (\vec{x})} \; \phi_2 (\vec{x}) \; \frac{dx\,dy}{y^2} \quad . \tag{B. 26}$$

Using (B. 25) one easily finds the infinitesimal generators of the representation U_ℓ^+ . The second order Casimir operator is:

$$\mathcal{C}_2 = y^2 \Delta - 2i\ell\, y \nabla_x \qquad (\Delta = \nabla_x^2 + \nabla_y^2) \; . \tag{B. 27}$$

Its eigenvalues are $\ell(\ell-1)$, $\ell = 1, 2, \ldots$, the corresponding eigenfunctions of the canonical basis have the form

$$\phi_{\ell k}(z) = A_{\ell k}\, u^\ell \left(\frac{\bar{z}-i}{z+i}\right)^\ell \left(\frac{z-i}{z+i}\right)^k , \quad u = \frac{4y}{x^2+(y+1)^2} , \quad k = 0, 1, 2, \ldots \tag{B. 28}$$

They are also eigenfunctions of the compact generator

$$X_{12} = \frac{d}{d\theta} U_\ell(k_\theta)\big|_{\theta=0} = -\left\{\tfrac{1}{2}(1+x^2-y^2)\nabla_x + xy\,\nabla_y + i\ell\, y\right\} \tag{B. 29}$$

(corresponding to eigenvalues $-i(\ell+k)$), the vectors $\phi \in \mathcal{H}_\ell$ are related to the analytic functions $f(z)$ (in $\mathbb{C}_+$) with boundary values $f(x)$ (in D_ℓ^+) by

$$\phi(z) = y^\ell f(z) \tag{B. 30}$$

In particular, the canonical basis functions (B. 28) go into

$$f_{\ell k}(z) = A_{\ell k} \left(\frac{4}{z^2+1}\right)^\ell \left(\frac{z-i}{z+i}\right)^{\ell+k} = 4^\ell A_{\ell k} \frac{(z-i)^k}{(z+i)^{2\ell+k}} \quad . \tag{B. 31}$$

The operator

$$L : \phi(z) \to f(x) = \lim_{y \downarrow 0} y^{-\ell} \phi(x+iy)$$

intertwines the representations U_ℓ^+ and T_ℓ^+ . Its inverse is given by the Cauchy integral formual: $L^{-1} : f(x) \to \phi(x) = y^\ell f(z)$ where

$$f(z) = \frac{1}{2\pi i} \int_{-\infty}^{\infty} \frac{f(x')\, dx'}{x' - z} \qquad (\operatorname{Im} z = y > 0) \ . \tag{B. 32}$$

It can also be related to the Green function

$$\mathcal{D}^+(z_1, z_2) = \frac{u^\ell}{4\pi} \Big(\sum_{k=1}^{2\ell-1} \frac{1}{ku^k} + \log \frac{1-u}{u} \Big) \Big(\frac{z_2 - \bar{z}_1}{z_1 - \bar{z}_2} \Big)^\ell ,$$

(B. 33)

$$u = \frac{(z_1 - \bar{z}_1)(\bar{z}_2 - z_2)}{|z_1 - \bar{z}_2|^2} = \frac{4 y_1 y_2}{(x_1 - x_2)^2 + (y_1 + y_2)^2}$$

for the eigenvalue equation of the Casimir operator

$$[y_1^2 \Delta_1 - 2i\ell\, y_1 \nabla_{x_1} - \ell\,(\ell - 1)] \mathcal{D}^+(z_1, z_2) = y_1^2\, \delta(x_1 - x_2)\, \delta(y_1 - y_2) \tag{B. 34}$$

To this end we write down the analogue of (7. 48), (7. 49) and take into account Eq. (B. 20) .

The scalar product (B. 26) in $\mathcal{H}_\ell$ is related to the invariant hermitian form on D_ℓ^+ as follows: Let $\phi_i(z) = y^\ell f_i(z)$, $i = 1, 2$; then inserting (B. 32) into (B. 26), we obtain:

$$(\phi_1, \phi_2) = \int \bar{f}_1(z) f_2(z) y^{2\ell-2} d^2z = \int dx_1 dx_2 d^2z \frac{\bar{f}_1(x_1) f_2(x_2)}{(x_1 - \bar{z})(x_2 - z)} y^{2\ell-2} =$$

$$= \frac{1}{2\pi} \int dx_1 \int dx_2 \bar{f}_1(x_1) \left(\frac{1}{4} x_{12}^2\right)^{\ell-1} \log \frac{4}{(x_{12}+i0)^2} f_2(x_2) \qquad \text{(B. 35)}$$

In deriving (B. 35) we used analytic regularization of the integral in y and the orthogonality of f_i to polynomials of degree smaller than $2\ell - 1$ (see (B. 20)).

APPENDIX C.

Positivity of the invariant scalar product in the subspace $D_{\ell\nu}$ of $C'^{+}_{\ell\nu}$

C.1 The problem. Asymptotic expansion of $f(p, \mathfrak{z})$ for $p \to 0$

The positivity of the expression (6.13) for $G'^{+}_{\ell\nu}(p)$ is not sufficient to conclude that the scalar product $(f, G'^{+}_{\ell\nu} f)$ is positive semidefinite in $C'^{-}_{\ell\nu}$, since the Fourier transforms $f(p)$ of elements of $C'^{-}_{\ell\nu}$ are in general singular for $p \to 0$ and the precise definition of the scalar product may require a regularization, which would destroy its positivity. The objective of this Appendix is to prove that this does not happen. The analysis is based on a study of the asymptotic behavior of $f(p)$ for $p \to 0$. Because of the isomorphism between the subspace $D_{\ell\nu}$ of $C'^{+}_{\ell\nu}$ and the factor space $C'^{-}_{\ell\nu}/F'_{\ell\nu}$ (see (6.19)) and the equivalence of the corresponding representations of G, this will be sufficient for establishing the unitarity of the representation $U_{\ell\nu}$ in $D_{\ell\nu}$.

The functions $f \in C_\chi$ are not in general integrable for Re $c < 0$ ($\chi = [\ell, c]$) and their Fourier transform does not exist in the L^1 sense. It can always be defined, however, away from the origin ($p = 0$) in momentum space and is a fast decreasing function of p for $p \to \infty$. (The last statement follows from the infinite differentiability of the functions $f(x) \in C_\chi$.) On the other hand, the asymptotic behavior of $f(p)$ for $p \to 0$ is determined from the asymptotic form (2.29) of $f(x)$ for $x \to \infty$.

If $c < 0$, then

$$f(p, \mathfrak{z}) \approx P_\ell(p, \mathfrak{z})\,(p^2)^c \tag{C.1}$$

where $P_\ell(p, \mathfrak{z})$ is a polynomial of p and $\mathfrak{z}$, homogeneous of degree ℓ in $\mathfrak{z}$. That follows from a general Paley-Wiener type theorem. In order to illustrate this property we shall evaluate the Fourier transform of functions $f \in C_\chi$ of the form

$$f(x, \mathfrak{z}) = (2\pi)^{-h} \left(\frac{2}{a^2+(x-b)^2}\right)^{h+c+\ell'} h_\ell(\{\mathfrak{z}, x\}), \qquad (C.2)$$

where h_ℓ is a homogeneous polynomial of degree ℓ of its (h+1)(2h+1) arguments $\{\mathfrak{z}, x\} = \{\mathfrak{z}, x^2 r(x)\mathfrak{z};\ x\mathfrak{z}, x \wedge \mathfrak{z}\}$, $\mathfrak{z} \in \mathbb{K}_{2h}$ (cf. (6.2b)), $a > 0$, and $\ell' \geq \ell$. We notice that the linear span $\mathcal{A}_\chi$ of functions of the form (C.2) is $O^\uparrow(2h+1,1)$ - invariant and dense in C_χ. The Fourier transform of f can be defined through analytic continuation in c from the right half plane (Re $c > 0$). Using Eqs. (3.915.5) and (6.565.4) of ref. [G7], we obtain

$$f(p, \mathfrak{z}) = h_\ell(\{\mathfrak{z}, i\nabla_p\})(2\pi)^{-h} \int \left(\frac{2}{a^2+(x-b)^2}\right)^{h+c+\ell'} e^{-ipx}\, dx$$

$$= 2^{h+c+\ell}\, h_\ell(\{\mathfrak{z}, i\nabla_p\})\, |p|^{1-h}\, e^{-ipb} \int_0^\infty \frac{r^h}{(a^2+r^2)^{h+c+\ell'}} J_{h-1}(|p|r)dr$$

$$= 2\, h_\ell(\{\mathfrak{z}, i\nabla_p\}) e^{-ipb} \frac{1}{\Gamma(h+c+\ell')} \left(\frac{|p|}{a}\right)^{c+\ell'} K_{c+\ell'}(a|p|). \qquad (C.3)$$

The right-hand side of (C.3) decreases exponentially for $|p| \to \infty$ as anticipated. For non-integer c the small p asymptotic behavior of $f(p,\mathfrak{z})$ is given by

$$f(p,\mathfrak{z}) \underset{p\to 0}{\approx} \frac{h_\ell(\{\mathfrak{z}, i\nabla_p\})}{\Gamma(h+c+\ell')} \left[\left(\frac{2}{a^2}\right)^{c+\ell'} \Gamma(c+\ell') + O(p) + \Gamma(-c-\ell')\left(\frac{p^2}{2}\right)^{c+\ell'} (1+O(p))\right] \tag{C.4}$$

($O(p)$ denotes as usual a quantity of the order of magnitude of p).

For negative (non-integer)c the second term is dominant. If $c+\ell'$ is a non-negative integer, then we have

$$f(p,\mathfrak{z}) \underset{p\to 0}{\approx} \frac{(-1)^{c+\ell'}}{\Gamma(h+c+\ell')} h_\ell(\{\mathfrak{z}, i\nabla_p\}) \left[\frac{1}{(c+\ell')!}\left(\frac{p^2}{2}\right)^{c+\ell'} \log\frac{4}{a^2p^2} \quad O(1)\right]. \tag{C.5a}$$

Finally, if $c+\ell'$ is a negative integer, we find

$$f(p,\mathfrak{z}) \underset{p\to 0}{\approx} \frac{h_\ell(\{\mathfrak{z}, i\nabla_p\})}{\Gamma(h+c+\ell')} (|c+\ell'|-1)! \left(\frac{p^2}{2}\right)^{c+\ell'} (1+O(p)). \tag{C.5b}$$

C.2 Existence of a non trivial positive semi-definite hermitian form $(f_1, G'^{+}_{\ell\nu} f_2)$ on $C'^{-}_{\ell\nu}$

Let now $\chi = \chi'^{-}_{\ell\nu} = [\ell+\nu, 1-\ell-h]$. Consider the quadratic form on $C'^{-}_{\ell\nu}$

$$(f, G'^{+}_{\ell\nu} f) = \lim_{\varepsilon\downarrow 0} \int_{p^2\geq\varepsilon} \langle f(p), G'^{+}_{\ell\nu}(p) f(p)\rangle \frac{d^{2h}p}{(2\pi)^{2h}} . \tag{C.6}$$

If we allow a priori the value $+\infty$, then the limit in the right-hand side always exists, since the integrand is non-negative. We remark that the set of f's for which this limit is finite forms a linear manifold. This follows from Schwartz's inequality for the integrand, which implies that

$$< f_1 + f_2, G'^{+}_{\ell\nu} (f_1 + f_2) > \;\leq 2 \left(< f_1, G'^{+}_{\ell\nu} f_1 > + < f_2, G'^{+}_{\ell\nu} f_2 > \right) .$$

The form (C. 6) vanishes for $f \in F'_{\ell\nu}$ ($\subset C'^{-}_{\ell\nu}$), since according to the results of Sec. 6,B the integrand in (C. 6) vanishes for such f's.

<u>Proposition C.1.</u> <u>The limit (C. 6) exists for every $f \in C'^{-}_{\ell\nu}$.</u> <u>It gives rise to a positive definite scalar product on the factor space</u> $C'^{-}_{\ell\nu}/F'_{\ell\nu}$.

<u>Proof.</u> Since $(f, G'^{+}_{\ell\nu} f) = 0$ for $f \in F'_{\ell\nu}$, it would be sufficient to prove the existence of a finite limit (C. 6) for a suitable representative in each coset in $C'^{-}_{\ell\nu}/F'_{\ell\nu}$. To each such coset we shall assign a representative f satisfying the p-space equation

$$(p\,\mathcal{D})^{\nu} f(p, z) = 0 \qquad \text{(C. 7)}$$

(where $\mathcal{D}$ is given by (A.47) (A.49). It is easily seen, that a representative with this property does exist and is determined uniquely by (C. 7). Indeed the Fourier transform of very element $f \in C'^{-}_{\ell\nu}$ can be written in the form $f = f_1 + f_2$, where

$$f_1(p) = \sum_{s=0}^{\ell} \prod^{\ell+\nu s}(p)\, f(p), \quad f_2(p) = \sum_{s=\ell+1}^{\ell+\nu} \prod^{\ell+\nu s}(p)\, f(p). \tag{C.8}$$

It follows from (5.21) (5.14) and (6.6) that $f_1 \in F'_{\ell\nu}$ and f_2 satisfies (C.7). On the other hand (5.21) also implies that if $f \in F'_{\ell\nu}$ satisfies (C.7) then $f = 0$.

Our next objective will be to replace the finite dimensional scalar product $\langle f(p), G'^{+}_{\ell\nu}(p) f(p)\rangle$ by a form suitable for exploiting (C.7).

Lemma C.2. Define the $M(=SO(2h))$ average of $f(p) \times \bar{f}(p)$ by

$$F(p; \zeta, \bar{\zeta}) = \int f(mp, m\zeta)\, \bar{f}(mp, m\bar{\zeta})\, dm, \tag{C.9}$$

where dm is the normalized Haar measure on M; then the angular integration in the right hand side of (C.6) can be expressed in the form

$$\int d\Omega_{\hat{p}} \left\langle f(p), G'^{+}_{\ell\nu}(p) f(p)\right\rangle = \sigma_{2h-1}\, \mathrm{tr}\,[G'^{+}_{\ell\nu}(p)\, F(p)] \tag{C.10a}$$

where σ_ν is the surface of the unit sphere, given by (A.21), and

$$\mathrm{tr}[G'^{+}_{\ell\nu}(p)\, F(p)] = [(\ell+\nu)!]^{-2}\, G'^{+}_{\ell\nu}(p; \partial_\zeta, \partial_{\bar{\zeta}})\, F(p; \zeta, \bar{\zeta}) \tag{C.10b}$$

is a function of p^2 only.

The lemma is a straightforward consequence of the rotation invariance of the integral over the unit sphere in momentum space and of the covariance property $G'^{+}_{\ell\nu}(mp; m\zeta_1, m\zeta_2) = G'^{+}_{\ell\nu}(p; \zeta_1, \zeta_2)$ of the intertwining kernel.

Eq. (C.1) implies that the M-invariant function F can be written in the form

$$F(p;\mathfrak{z},\bar{\mathfrak{z}}) = \phi(p^2\Pi, \mathfrak{z}\bar{\mathfrak{z}}; p^2)(p^2)^{2(1-\ell-h)}, \qquad p^2\Pi = p^2\mathfrak{z}\bar{\mathfrak{z}} - (p\mathfrak{z})(p\bar{\mathfrak{z}}) \tag{C.11}$$

where ϕ is a polynomial in all three arguments, homogeneous of degree $\ell+\nu$ with respect to the first two of them. Furthermore according to (C.7), it satisfies the equation

$$(pD)^{\nu}\phi(p^2\Pi, \mathfrak{z}\bar{\mathfrak{z}}; p^2) = 0 \tag{C.12}$$

where $(pD)^{\nu}$ is given by (A.55), (A.60).

We shall show that any polynomial solution of (C.12) is bounded by $C(p^2)^{\ell+1}$ for $p \to 0$. Indeed every polynomial ϕ, satisfying the above homogeneity conditions, can be written in the form

$$\phi(p^2\mathfrak{z}\bar{\mathfrak{z}} - (p\mathfrak{z})(p\bar{\mathfrak{z}}), \mathfrak{z}\bar{\mathfrak{z}}; p^2) = (\mathfrak{z}\bar{\mathfrak{z}})^{\nu}\phi_{\ell}(p^2\Pi, \mathfrak{z}\bar{\mathfrak{z}}; p^2) + \\ + \phi_1(p^2\Pi, \mathfrak{z}\bar{\mathfrak{z}}; p^2), \tag{C.13}$$

where

$$| \phi_1 (p^2\Pi , \mathfrak{z}\bar{\mathfrak{z}} ; p^2)| \leq C (p^2)^{\ell+1} \tag{C.14}$$

and ϕ_ℓ is a homogeneous polynomial degree ℓ of the first two arguments and a polynomial p of overall degree not exceeding 2ℓ. Since Eq. (C.12) is satisfied identically with respect to p, it has to be satisfied by each of the two terms in the right hand side of (C.13).

Lemma C.3. The only solution of the equation

$$(p\cdot D)^\nu \{ (\mathfrak{z}\bar{\mathfrak{z}})^\nu \phi_\ell (p^2\Pi , \mathfrak{z}\bar{\mathfrak{z}} ; p^2)\} = 0 \tag{C.15}$$

satisfying the above conditions is $\phi_\ell = 0$.

If Lemma C.3 is proven, then the validity of Proposition C.1 will follow from the behavior of $G'^{+}_{\ell\nu}(p) \sim (p^2)^{\ell+h-1}$ for small p and from (C.10), (C.11), (C.13), and (C.14).

Proof of Lemma C.3. Assume that there is a non-zero polynomial ϕ_ℓ (of degree ℓ) satisfying (C.15). Let s be the maximal non-negative integer for which the coefficient $a_s(p^2)$ in the expansion

$$(\mathfrak{z}\bar{\mathfrak{z}})^\nu \phi_\ell (p^2\Pi , \mathfrak{z}\bar{\mathfrak{z}} ; p^2) = \sum_{k=0}^{s} a_k(p^2) (p^2\Pi)^k (\mathfrak{z}\bar{\mathfrak{z}})^{\ell+\nu-k} \tag{C.16}$$

does not vanish. Here $a_k(p^2)$ is a polynomial of degree $\ell-k$ and $s \leq \ell$. Inserting (C.16) in (C.12) and taking into account the equations:

$$p\partial_{\mathfrak{z}} \Pi(p ; \mathfrak{z}, \bar{\mathfrak{z}}) = 0 \;, \quad (p\partial_{\mathfrak{z}})^i (\mathfrak{z}\bar{\mathfrak{z}})^j = \frac{j!}{(j-i)!} (p\bar{\mathfrak{z}})^i (\mathfrak{z}\bar{\mathfrak{z}})^{j-i} \;, \quad \Delta_{\mathfrak{z}} (\mathfrak{z}\bar{\mathfrak{z}})^j = 0$$

$$\Delta_{\mathfrak{z}}^i (p\Pi)^j = (-1)^i \frac{j!}{(j-2i)!} (p^2)^i (p\bar{\mathfrak{z}})^{2i} (p^2\Pi)^{j-2i}$$

$$(\partial_\mu (p^2\Pi)^i)\, \partial_\mu (\mathfrak{z}\bar{\mathfrak{z}})^j = -ij\, (p^2\Pi)^{i-1} (\mathfrak{z}\bar{\mathfrak{z}})^{j-1} (p\bar{\mathfrak{z}})^2 \;, \quad \partial_\mu \equiv \frac{\partial}{\partial \mathfrak{z}_\mu} \tag{C.17}$$

$$\Delta^m (p^2\Pi)^i (\mathfrak{z}\bar{\mathfrak{z}})^j = \sum_{k=0}^{m} \binom{m}{k} 2^{m-k} (\Delta^k \partial_{\mu_1} \cdots \partial_{\mu_{m-k}} (p^2\Pi)^i)\, \partial_{\mu_1} \cdots \partial_{\mu_{m-k}} (\mathfrak{z}\bar{\mathfrak{z}})^j$$

we obtain $(pD)^{\nu}(\mathfrak{z}\bar{\mathfrak{z}})^{\nu}\,\phi_{\ell}\,(p^2\Pi,\mathfrak{z}\bar{\mathfrak{z}};p^2) =$

$$= \sum_{\kappa=0}^{s} a_{\kappa}(p^2)\,\kappa!\,(\ell-\nu-\kappa)!\,(p\bar{\mathfrak{z}})^{\nu}\sum_{i}\left(\frac{p^2}{2}\right)^{i}(p^2\Pi)^{\kappa-i}(\mathfrak{z}\bar{\mathfrak{z}})^{\ell-\kappa+i}\,\cdot$$

$$\cdot\sum_{mnj}\frac{(-1)^{m+n+j}\nu!\,(h+\ell-1)_{\nu-m}\,m!\,2^{j}}{(\nu-m-n)!\,n!\,(i-n-j)!\,(m-i+n+j)!}\cdot\frac{1}{(\kappa-m-i+n+j)!\,(\ell-\kappa+i-j)!\,(m-n-j)!\,j!} \qquad \text{(C.18)}$$

$$= a_s(p^2)\,(p\bar{\mathfrak{z}})^{\nu}(p^2\Pi)^{s}\,(\mathfrak{z}\bar{\mathfrak{z}})^{\ell-s}\frac{(\ell+\nu-s)!}{(\ell-s)!}\sum_{m}\frac{(-1)^{m}\nu!\,(h+\ell-1)_{\nu-m}\,s!}{(\nu-m)!\,(s-m)!\,m!} + (p\bar{\mathfrak{z}})^{\nu}(p^2\Pi)^{s-1}(\mathfrak{z}\bar{\mathfrak{z}})^{\ell-s+1}\mathcal{B}_{\ell\nu s}(p^2) + \ldots = 0$$

For the sum in the first term we have (see Appendix E below, Eq. (E.10)

$$\sum_{m}\frac{(-1)^{m}\,\nu!\,s!\,(h+\ell-1)_{\nu-m}}{(\nu-m)!\,(s-m)!\,m!} = (h+\ell-s-1)_{\nu} \qquad \text{(C.19)}$$

Thus the coefficient of $a_s(p^2)(p^2\Pi)^{s}$ is not zero, which implies

$a_s(p^2) = 0$, contrary to our assumption.

PART TWO

CONFORMAL PARTIAL WAVE ANALYSIS

SYNOPSIS

We study the tensor product decomposition of two unitary irreducible representations of the Euclidean conformal group, $O^{+}(2h+1,1)$. Conditions are found under which only principal series representations contribute to the decomposition. Clebsch Gordan kernels are evaluated which satisfy a completeness and orthogonality relation. Convenient normalization conventions are discussed for Clebsch Gordan kernels which effect the decomposition of the tensor product of two class I ("scalar") representations, and identities are found which are satisfied by their analytical continuation at partially equivalent integer points.

The results are applied to write down conformal partial wave expansions for 1-particle irreducible n-point Green functions in conformal invariant quantum field theory.

Special cases of these expansions are used to derive an expansion of the type

$$\langle \varphi(x_1)\dots\varphi(x_n)\rangle_0 = \langle \varphi(x_1)\varphi(x_2)\rangle_0 \langle \varphi(x_3)\dots\varphi(x_n)\rangle_0 + \\ + \sum_{\chi_\ell} C^2(\chi_\ell)\int (dp)\, Q^{\tilde{\chi}_\ell}(x_1 x_2; -p)\, w_{\chi_\ell}(p)\, Q^{\tilde{\chi}_\ell}(p; x_3 \dots x_n) \,. \qquad (0)$$

Here $\chi_\ell = [\ell, c_\ell]$ are labels for infinite dimensional symmetric tensor representations of the Euclidean conformal group, $\tilde{\chi}_\ell = [\ell, -c_\ell]$, the constants $C(\chi_\ell)$ are real, and $Q^{\tilde{\chi}}$ and w_χ have the properties of vacuum expectation values of field products. The starting point is an infinite set of coupled non-linear integral equations for Euclidean Green functions in 2h space-time dimensions of the type written some 15 years ago by Fradkin and Symanzik. The Green functions of the corresponding Gell-Mann - Low limit theory are expanded in conformal partial waves. The dynamical equations imply the existence of poles and factorization of residues in the partial waves as functions of the representation parameters. In proving the validity of (O) we use some differential relations between partially equivalent exeptional representations of $O^{+}(2h+1,1)$, established in Part One.

IV. CLEBSCH-GORDAN EXPANSION OF THE TENSOR PRODUCT OF TWO UNITARY PRINCIPAL OR SUPPLEMENTARY SERIES REPRESENTATIONS

9. The Kronecker product of two elementary representations as an induced representation on G/MA *)

The Kronecker product $\chi_1 \otimes \chi_2$ of two elementary representations $\chi_1 = [\ell_1, c_1]$ and $\chi_2 = [\ell_2, c_2]$ of G acts in the space of infinitely differentiable functions $f(g_1, g_2)$ on $G \times G$ with values in the vector space $V^{\ell_1} \otimes V^{\ell_2}$. Considered as functions of the individual variables they share the covariance properties of functions in the elementary representation spaces $\mathcal{C}_{\chi_1}$ resp. $\mathcal{C}_{\chi_2}$, cp. Sec. 2.B. In particular,

$$f(g_1 p_1, g_2 p_2) = [D^{\chi_1}(p_1^{-1}) \otimes D^{\chi_2}(p_2^{-1})] f(g_1, g_2) \qquad \text{for } p_1, p_2 \text{ in MAN.} \tag{9.1}$$

The action of G on such functions is given by

$$(T(g) f)(g_1, g_2) = f(g^{-1} g_1, g^{-1} g_2) \tag{9.2a}$$

Because of covariance property (9.1), functions f are uniquely determined by their restriction $f(x_1 x_2) = f(\tilde{n}_{x_1}, \tilde{n}_{x_2})$ $(x_1 \neq x_2)$ to $\tilde{N} \times \tilde{N}$. Transformation law (9.2a) reads then

$$(T(g) f)(x_1 x_2) = [D^{\chi_1}(p(x_1, g)) \otimes D^{\chi_2}(p(x_2, g))] f(g^{-1} x_1, g^{-1} x_2) \,. \tag{9.2b}$$

The cocycles $p(x,g) \in$ MAN are defined by (1.27a), viz. $g^{-1} \tilde{n}_x = \tilde{n}_{g^{-1}x}\, p(x,g)^{-1}$.

If χ_1, χ_2 belong to the unitary principal series then also their Kronecker product is unitary by virtue of the inherited G-invariant scalar product

$$(f_1, f_2) = \int dx_1 dx_2 < f_1(x_1 x_2), f_2(x_1 x_2) > \tag{9.3}$$

$<,>$ is the M-invariant scalar product on $V^{\ell_1} \otimes V^{\ell_2}$. All this parallels the discussion of elementary representations in the "noncompact picture" in Sec. 2.D.

We will now demonstrate that the Kronecker product $\chi_1 \otimes \chi_2$ may also be regarded as an induced representation on G/MA.

We define a map Q of $\mathcal{C}_{\chi_1} \otimes \mathcal{C}_{\chi_2}$ into a space of infinitely differentiable functions on G with values in $V^{\ell_1} \otimes V^{\tilde{\ell}_2}$ and covariance property

$$F(gma) = L(ma)^{-1} F(g) \qquad \text{with} \quad L(ma) = |a|^{c_2 - c_1} [D^{\ell_1}(m) \otimes D^{\tilde{\ell}_2}(m)] \,. \tag{9.4}$$

---

*) In Secs. 9 and 10 we use a different parametrization of the subgroup N. What was formerly called n_x is now $n_{\theta x}$, θ = reflection of the 2h-th axis. With the new notation we have $n_x = w\tilde{n}_x w$ (while with the notation of part one we had $R\tilde{n}_x R = n_{-x}$, $w\tilde{n}_x w = n_{\theta x}$).

for $ma \in MA$. The map Q is defined by

$$(Qf)(g) = f(g, gw) \tag{9.5}$$

w is the Weyl-inversion $wx = \theta x/x^2$; it belongs to the identity component of G as we know. We recall that $m^w \equiv wmw^{-1} = \theta m \theta \in M$ for $m \in M$, and $waw^{-1} = a^{-1}$ for a in A. Condition (9. 4) is then an immediate consequence of the definition (9. 5) and the covariance property (9.1) of f ; viz. $f(gma, gmaw) = f(gma, g^w m^w a^{-1}) =$
$= [D^{\chi_1}(ma)^{-1} \otimes D^{\chi_2}(m^w a^{-1})^{-1}] f(g, g^w)$.
Since $D^{\chi}(ma) = |a|^{-h-c} D^{\ell}(m)$ by definition (2.28) we obtain indeed (9. 4).

The space $Q(\mathcal{E}_{\chi_1} \otimes \mathcal{E}_{\chi_2})$ of functions with covariance property (9. 4) carries a representation of G ,

$$(\mathcal{T}(g)F)(g') = F(g^{-1}g'). \tag{9.6}$$

It is immediate from the definition (9. 5) of Q that $\mathcal{T}(g)Qf = Q\mathcal{T}(g)f$. I.e. Q has the intertwining property.

We will show that functions f in $\mathcal{E}_{\chi_1} \otimes \mathcal{E}_{\chi_2}$ are uniquely determined by Qf .

Lemma 9.1: The group G acts transitively on noncoinciding pairs (x_1, x_2), $x_1 \neq x_2$ of points in Euclidean space, with subgroup of stability isomorphic to MA.

To prove this crucial lemma, consider first the special pair $(\dot{x}_1 = 0, \dot{x}_2 = \infty)$. Its subgroup of stability in G is MA. On the other hand

$$\tilde{n}_x n_{wy}(0, \infty) = (x_1, x_2) \quad \text{for} \quad x = x_1, \; y = x_2 - x_1. \tag{9.7a}$$

Indeed $\tilde{n}_x n_{wy}(0,\infty) = \tilde{n}_x w \tilde{n}_{wy} w(0,\infty) = \tilde{n}_x w \tilde{n}_{wy}(\infty, 0) =$
$= \tilde{n}_x w(\infty, wy) = \tilde{n}_x(0, y) = (x, x+y) = (x_1, x_2)$.

So every point (x_1, x_2) with $x_1 \neq x_2$ may be reached from $(0,\infty)$ by applying a suitable conformal transformation. This proves transitivity .

We remark that also

$$n_{wy'} \tilde{n}_{x'}(0,\infty) = (x_1, x_2) \text{ for } y' = x_2, \; x' = w(wx_1 - wx_2). \tag{9.7b}$$

The connection is given by the following corollary of identities (1.27a) and (1.27f):

$$\tilde{n}_x n_{wy} = n_{wy'} \tilde{n}_{x'} h \quad \text{with} \quad h = h_{y'} h_y^{-1} = h_{x'} h_x^{-1} \in AM \text{, and} \quad y' = y+x \text{, } x' = w(wx - wy') \tag{9.8}$$

Explicitly $h_z = m_z a_z$ with $|a_z| = z^2$, $m_z = -r(z)\theta$, $r(z)^\mu{}_\nu = -\delta^\mu{}_\nu + 2z^\mu z_\nu / z^2$.

Now we want to reconstruct f from Qf. It suffices to determine $f(x_1 x_2) = f(\tilde{n}_{x_1}, \tilde{n}_{x_2})$ for $x_1 \neq x_2$. Pairs (g_1, g_2) which cannot be written in the form $(\tilde{n}_{x_1} p_1, \tilde{n}_{x_2} p_2)$ with p_1, p_2 in MAN and $x_1 \neq x_2$ form a lower dimensional submanifold of $G \times G$; and $f(g_1, g_2)$ was assumed infinitely differentiable. Consider then

$$(Qf)(\tilde{n}_x n_{wy}) = f(\tilde{n}_x n_{wy}, \tilde{n}_x n_{wy} w)$$

One has $n_{wy} w = w \tilde{n}_{wy} = \tilde{n}_y p(wy, w)^{-1} = \tilde{n}_y p_{wy}^{-1}$ with $p_{wy} \in P$ and given explicitly by (1.27e). Covariance property (9.1) gives then

$$F(\tilde{n}_x n_{wy}) \equiv (Qf)(\tilde{n}_x n_{wy}) = [\mathbb{1} \otimes D^{\chi_2}(p_{wy})] f(x_1 x_2) \quad \text{with} \quad x_1 = x \text{, } x_2 = y + x \text{.} \tag{9.9a}$$

The matrix in the brackets has an inverse $\mathbb{1} \otimes D^{\chi_2}(p_{wy})^{-1}$, and so f is uniquely determined by Qf. In other words, we can define an inverse Q^* to Q by

$$f(x_1 x_2) = (Q^* F)(\tilde{n}_{x_1}, \tilde{n}_{x_2}) = [\mathbb{1} \otimes D^{\chi_2}(p_{wy})^{-1}] F(\tilde{n}_x n_{wy}) \tag{9.9b}$$

If one writes $p_{wy} = h_{wy} n_{wy}$ then $D^{\chi_2}(p_{wy}) = D^{\chi_2}(h_{wy})$ by definition.

There is an equivalent formula which is based on (9.7b). It is derived from (9.9a) by exploiting (9.8):

$$(Qf)(n_{wy'} \tilde{n}_{x'}) = (Qf)(\tilde{n}_x n_{wy} h^{-1}) = L(h)(Qf)(\tilde{n}_x n_{wy})$$

by covariance property (9.4). Since $L(h) = D^{\chi_1}(h) \otimes D^{\chi_2}(h^w)$ one has $L(h)[\mathbb{1} \otimes D^{\chi_2}(p_{wy})] = D^{\chi_1}(h) \otimes D^{\chi_2}(h^w h_{wy})$. By (9.8), $h^w = h_{wy'} h_{wy}^{-1}$, so $D^{\chi_2}(h^w h_{wy}) = D^{\chi_2}(h_{wy'})$. Altogether

$$(Qf)(n_{wy'} \tilde{n}_{x'}) = [D^{\chi_1}(h) \otimes D^{\chi_2}(h_{wy'})] f(x_1 x_2) \tag{9.9c}$$

with $y' = x_2$, $x' = w(wx_1 - wx_2)$ and $h = h_{y'} h_{wy} = h_{x'} h_{wx'}$; $x = x_1$, $y = x_2 - x_1$.

We consider now the special case that $L(ma)$ is unitary. This is true if $c_2 - c_1$ is pure imaginary. This includes the case when χ_1 and χ_2 belong to the unitary principal series, i.e. c_1 and c_2 are pure imaginary. A G-invariant scalar product may then be defined by

$$(F_1, F_2) = \iint_{N \tilde{N}} dn\, d\tilde{n} \,\langle F_1(\tilde{n}n), F_2(\tilde{n}n)\rangle, \qquad (9.10)$$

We show that this agrees with the scalar product (9. 3) for the Kronecker product $\chi_1 \otimes \chi_2$ in case that χ_1 and χ_2 belong to the unitary principal series, viz.

$$(Qf_1, Qf_2) = (f_1, f_2) \qquad (9.11)$$

where the l.h.s. is defined by (9.10) and the r.h.s. by (9. 3). One speaks of a "unitary induced representation on G/MA" if the inducing representation L(ma) is unitary <u>and</u> the scalar product is defined by (9.10). [Note that $dn\,d\tilde{n}$ is an invariant measure on G/MA, and functions F are specified by their values on $N\tilde{N} \approx$ G/MA.]

<u>Theorem</u> 9.2. The Kronecker product of two representations χ_1 and χ_2 of the unitary principal series is unitarily equivalent to a unitarily induced representation on G/MA with inducing representation L of MA given by (9. 4).

<u>Proof</u>: The equivalence is effected by the map Q . It only remains to prove (9.11).

$(Qf_1, Qf_2) = \int dx\,dy \langle f_1(\tilde{n}_y n_x, \tilde{n}_y n_x w)\, f_2(\tilde{n}_y n_x, \tilde{n}_y n_x w)\rangle$. We write $\tilde{n}_y n_x w =$

We have, using (4.13), $\tilde{n}_y n_x w = \tilde{n}_y w \tilde{n}_x = \tilde{n}_y \tilde{n}_{wx} n_{-x} a_{wx} I_s r(x)$

Also $dx = (x^2)^{2h} dz = |a_x|^{2h} dz$, where $p_x = m_x a_x n$ as usual.

If all this is inserted and use is made of the covariance condition (9.I), equality (9.11) is readily established.

We note also that the inducing representation $L(ma)$ of MA is not irreducible unless $\ell_1 = 0$ or $\ell_2 = 0$ the trivial representation of M. As a first step towards the decomposition of the induced representation one can decompose L into irreducibles. We write $V^{\tilde{\ell}_2}$ in place of V^{ℓ_2} if we want to consider this vector space as carrier of the representation $D^{\tilde{\ell}_2}(m) = D^{\ell_2}(\theta m \theta)$ of M. Let us decompose

$$V^{\ell_1} \otimes V^{\tilde{\ell}_2} = \sum_j V^j \quad \text{, sum over } j \in \hat{M} \text{ such that } j \subset \ell_1 \otimes \tilde{\ell}_2 .$$

Define Clebsch Gordan maps $C(\ell_1 \tilde{\ell}_2 ; j)$ and their adjoints, the M-invariant imbeddings $C(\ell_1 \tilde{\ell}_2 ; j)^*$

$$C(\ell_1 \tilde{\ell}_2 ; j) : V^{\ell_1} \otimes V^{\tilde{\ell}_2} \mapsto V^j \quad ; \quad C(\ell_1 \tilde{\ell}_2 ; j)^* : V^j \mapsto V^{\ell_1} \otimes V^{\tilde{\ell}_2}.$$

Then $\sum_j C(\ell_1 \tilde{\ell}_2 ; j)^* C(\ell_1 \tilde{\ell}_2 ; j) = \text{id}$; and

$$C(\ell_1 \tilde{\ell}_2 ; j)[D^{\ell_1}(m) \otimes D^{\tilde{\ell}_2}(m)] = D^j(m)\, C(\ell_1 \tilde{\ell}_2 ; j). \tag{9.12a}$$

We may then decompose

$$F(g) = \sum_j C(\ell_1 \tilde{\ell}_2 ; j)^* F^j(g) \quad ; \quad F^j(g) = C(\ell_1 \tilde{\ell}_2 ; j) F(g) \in V^j \tag{9.12b}$$

and covariance condition (9. 4) becomes, for $ma \in MA$

$$F^j(gma) = L^j(ma)^{-1} F^j(g)$$

with

$$L^j(ma) = |a|^{c_2 - c_1} D^j(m). \tag{9.13}$$

It will be convenient to introduce the map

$$Q_j = C(\ell_1 \tilde{\ell}_2 ; j) \circ Q \tag{9.14}$$

and $Q_j^* = Q^* \circ C(\ell_1 \tilde{\ell}_2 ; j)^*$, cp. (9. 9b). If

$$F^j = Q_j f \quad \text{then} \quad f = \sum Q_j^* F^j, \tag{9.15}$$

since Q^* is the inverse of Q and the CG-map $C(\ell_1 \tilde{\ell}_2 ; j)$ is unitary.

The sum runs over UIR's j of M such that $j \subset \ell_1 \otimes \tilde{\ell}_2$.
If χ_1 and χ_2 are in the unitary principal series then Q_j^* is the adjoint of the operator Q_j .

10. Construction of the Clebsch Gordan expansion

10.A Clebsch Gordan kernels

We seek G-invariant maps

$$V : \chi_1 \otimes \chi_2 \longmapsto \chi \qquad (10.1)$$

from the Kronecker product of two elementary representations χ_1, χ_2 to another elementary representation χ. Since all concerned are function spaces, the maps will be effected by integral kernels. If $\chi_1 = [\ell_1, c_1]$, $\chi_2 = [\ell_2, c_2]$ and $\chi = [\ell, c]$ then $\chi_1 \otimes \chi_2$ is made up of functions with values in $V^{\ell_1} \otimes V^{\ell_2}$ while χ consists of functions with values in the finite dimensional representation space V^{ℓ} of M. In physical applications one works in the "noncompact picture", i.e. with functions on x-space. The kernels are then maps (= matrices)

$$V(x_1\tilde{\chi}_1\, x_2\tilde{\chi}_2 \,;\, x\chi) : V^{\ell_1} \otimes V^{\ell_2} \longmapsto V^{\ell} \qquad (10.2)$$

with covariance property

$$D^{\chi}(p_3) V(x_1'\tilde{\chi}_1 x_2'\tilde{\chi}_2 ; x_3'\chi)\,[D^{\tilde{\chi}_1}(p_1) \otimes D^{\tilde{\chi}_2}(p_2)] = V(x_1\tilde{\chi}_1 x_2\tilde{\chi}_2 ; x_3\chi)$$

$$\text{for } x_i' = g^{-1}x_i \,,\; p_i = p(x_i, g) \quad , g \text{ in } G.$$

As is usual in representation theory, we start by defining the map V on infinitely differentiable functions. We are looking for a complete set V^{js} of maps (10.1). The meaning of the labels j, s will emerge later.

We make use of the result of the preceding subsection. Let $\sigma = c_1 - c_2$ and $\Xi^{j\sigma}$ be the induced representation on G/MA which is induced by the representation L^j of MA and consists of functions $F^j = Q_j f$ (f in $\mathcal{E}_{\chi_1} \otimes \mathcal{E}_{\chi_2}$) on G with covariance property (9.3).

Let

$$t^{js} : \Xi^{j\sigma} \longmapsto \chi \qquad (10.3)$$

a complete set (labelled by s) of G-invariant maps from $\Xi^{j\sigma}$ to the elementary representation χ.

Then a complete set of maps (10.1) is given by

$$v^{js} = t^{js} \circ Q_j \quad : \quad \chi_1 \otimes \chi_2 \xrightarrow{Q_j} \Xi^{j\sigma} \xrightarrow{t^{js}} \chi \qquad (10.4)$$

since $\chi_1 \otimes \chi_2 = \sum_j Q_j^* \Xi^{j\sigma}$ by (10.1).

So we have to find the maps t^{js} first. They are intertwining (i.e. G-invariant) maps between two induced representations. As such they can be found by a standard procedure due to Bruhat.

Functions in $\Xi^{j\sigma}$ admit an integral representation

$$F^j(g) = \int_{MA} dm\,da\; L^j(ma)\, \tilde{F}^j(gma). \qquad (10.5)$$

This makes covariance property (9.23) manifest for arbitrary $\tilde{F}$. Of course $\tilde{F}$ is not uniquely determined by F.
The map t^{js} will be implemented by an integral kernel $t^{js}(g_1, g_2)$, $g_i \in G$,viz.

$$(t^{js} F)(g_1) = \int_G dg_2\, t^{js}(g_1, g_2)\, \tilde{F}(g_2). \qquad (10.6)$$

It will be required that this be independent of the choice of $\tilde{F}$ provided it satisfies (10.5). In particular let h in MA. Then $\tilde{F}$ and $\tilde{F}'(g) = L(h)\tilde{F}(gh)$ determine the same F. Therefore, we must have $\int dg_2\, t^{js}(g_1, g_2)\tilde{F}(g_2) =$
$= \int dg_2\, t^{js}(g_1, g_2) L^j(h)\tilde{F}(g_2 h) = \int dg_2'\, t^{js}(g_1, g_2' h^{-1}) L(h)\tilde{F}(g_2')$.
Since $\tilde{F}$ is arbitrary, this requires

$$t^{js}(g_1, g_2 h^{-1}) L^j(h) = t^{js}(g_1, g_2) \quad \text{for } h \text{ in MA}. \qquad (10.7a)$$

Conversely, covariance property (10.7a) makes it possible to rewrite the r.h.s. of (10.6) in terms of F itself. This will be exploited later on, cp. Eq. (10.29) below. For now we continue with the determination of kernels $t^{js}(g_1, g_2)$. We have to meet two more requirements beyond (10.7a).

i) The map t^{js} must be G-invariant.

ii) $t^{js}F$ must belong to χ for F in $\Xi^{j\sigma}$, so it must have covariance property $(t^{js}F)(gp) = D^{\chi}(p)^{-1}(t^{js}F)(g)$ for p in MAN, or, using (10.6),

$$\int dg_2\, t^{js}(g_1 p, g_2)\tilde{F}(g_2) = \int dg_2\, D^{\chi}(p)^{-1} t^{js}(g_1, g_2)\tilde{F}(g_2) .$$

Since $\tilde{F}$ is arbitrary we deduce the covariance requirement

$$t^{js}(g_1 p, g_2) = D^{\chi}(p)^{-1} t^{js}(g_1, g_2) \quad \text{for } p \in \text{MAN}. \qquad (10.7\text{b})$$

Condition i), viz. G-invariance of the map t^{js}, means that $\mathcal{T}(g)t^{js}F(g_1) = t^{js}\mathcal{T}'(g)F(g_1)$ The l.h.s. is $(t^{js}F)(g^{-1}g_1)$ by transformation law (2.15) for elementary representations. Concerning the r.h.s. we note that $(\mathcal{T}'(g)F)(g_2)$ $= F(g^{-1}g_2) = \int dm\,da\, L^j(ma)\tilde{F}(g^{-1}g_2 ma)$. Equality of both sides reads then

$$\int dg_2\, t^{js}(g^{-1}g_1, g_2)\tilde{F}(g_2) = \int dg_2\, t^{js}(g_1, g_2) L^j(ma)\tilde{F}(g^{-1}g_2 ma) = \int dg_2'\, t^{js}(g_1, gg_2')\tilde{F}(g_2').$$

The last equality is based on G-invariance of the Haar measure dg_2 . Since $\tilde{F}$ is arbitrary

$$t^{js}(gg_1, gg_2) = t^{js}(g_1, g_2) \quad \text{for } g, g_1, g_2 \text{ in G}. \qquad (10.7\text{c})$$

Our problem is to find the most general solution of Eqs. (10.7 a,b,c).

The most general solution of (10.7c) is

$$t^{js}(g_1, g_2) = t_{\#}^{js}(g_2^{-1}g_1) \qquad (10.8)$$

where $t_{\#}^{js}$ are matrix-valued generalized functions of one variable g in G. Covariance conditions (10.7b,c) become in terms of $t_{\#}$

$$t_{\#}^{js}(hgp^{-1}) = D^{\chi}(p)t_{\#}^{js}(g)L^j(h)^{-1} \quad \text{for } h \in H = MA,\ p \in P = \text{MAN} \qquad (10.9)$$

Here and in the following we abbreviate the groups MA = H, MAN = P. We have to find the most general solution of covariance condition (10.9).

Let us define an action of the group $H \times P$ on the

manifold G by

$$(h, p)g = hgp^{-1}. \qquad (10.10)$$

This satisfies the group's composition law $(h_1, p_1)(h_2, p_2) = (h_1h_2, p_1p_2)$.
We may therefore decompose G into orbits on which $H \times P$ acts transitively.

We show that there are three orbits P, wP, and the remainder G' of G. G' is the only open orbit, the others have lower dimension.

It is clear that P and wP are orbits; for they are invariant (e.g. $hwPp^{-1} = wh^wP = wP$ since $h^w = whw^{-1} \in H \subset P$) and right multiplication with P alone acts transitively already. It remains to be shown that $H \times P$ acts transitively on G'.
G' is union of cosets $\tilde{n}_x P$ with $x \neq 0, \infty$. Right multiplication with P acts transitively on individual cosets. Left multiplication with H gives $h\tilde{n}_x P = \tilde{n}_{hx} P$. H = MA acts transitively on the pointed x-space $\mathbb{R}^{2h} \setminus \{0\}$. Therefore hx is arbitrary nonzero for any given $x \neq 0$. This proves transitivity.
Corresponding with the three orbits there may be three classes of invariant kernels with support concentrated on P, wP and the closure $\overline{G'}$ = G of G' respectively. We shall want to extend the map V to measurable functions later on (when we deal with Kronecker product of principal series representations). Kernels concentrated on P or wP cannot be used then* and it suffices to know the kernel $t_{\#}(g)$ for g in G'.

---

*They can however be found by the Bruhat method also, cp.[W3]. Alternatively one can use a physicists method which works in practice and which will be implicit in our later arguments. In the square integrable case (principal series) the singular functions on G' specify invariant distributions defined by an ordinary integral. These distributions are meromorphic in the continuous parameters c, c_1, c_2, and the invariant kernels with support on P or wP appear as residues at the poles in these parameters (if the appropriate normalization is chosen).

We determine the little group of G' in $H \times P$. We choose a standard point $\tilde{n}_{-\hat{x}}$, $\hat{x} = (\underset{\sim}{0}, 1)$. Let the "rotation group"

$$\mathcal{U} \subset M \quad \text{consist of } u \text{ in } M \text{ such that } u\hat{x} = \hat{x} . \tag{10.11}$$

Certainly $h\tilde{n}_{-\hat{x}}p^{-1} = \tilde{n}_{-\hat{x}}$ is true for $(h,p) = (u,u)$, $u \in \mathcal{U}$

Conversely $h\tilde{n}_{-\hat{x}}p^{-1}P = -h\hat{x}$ is equal to $\tilde{n}_{-\hat{x}}P = -\hat{x}$

only if $h \in \mathcal{U}$, and $u\tilde{n}_{-\hat{x}}u^{-1}p'^{-1} = \tilde{n}_{-\hat{x}}$ only if $p' = e$

Therefore $p' = pu = u$ if $h = u$.

In conclusion, the little group of $\tilde{n}_{-\hat{x}} \in G'$ consists of pairs

$$(u, u) \in H \times P \quad \text{with} \quad u \in \mathcal{U} . \tag{10.12}$$

Because of transitivity of $H \times P$ on G', every g in G' (and therefore almost every g in G) may be written in the form

$$g = h\tilde{n}_{-\hat{x}}p^{-1} \qquad \text{with } h \text{ in } H , \ p \text{ in } P . \tag{10.13}$$

Covariance condition (10.9) says that

$$t_{\#}^{js}(h\tilde{n}_{-\hat{x}}p^{-1}) = \mathcal{D}^{\chi}(p)\hat{t}^{js}L^{j}(h)^{-1} \quad \text{with } \hat{t}^{js} = t_{\#}^{js}(\tilde{n}_{-\hat{x}}) \tag{10.14a}$$

For consistency, $\hat{t}^{js}$ must be $\mathcal{U}$-invariant

$$\hat{t}^{js} = \mathcal{D}^{\chi}(u)\hat{t}^{js}L^{j}(u)^{-1} \equiv \mathcal{D}^{\ell}(u)\hat{t}^{js}\mathcal{D}^{j}(u)^{-1} \qquad \text{for } u \text{ in } \mathcal{U} . \tag{10.14b}$$

Conversely, given $\hat{t}^{js}$ which satisfies (10.14b), a kernel $t_{\#}^{js}(g)$ is defined by (10.14a) which in turn gives rise to an intertwining map V^{js} in manner explained above.

We classify matrices $\hat{t}^{js}$; they are $\mathcal{U}$-invariant maps $\mathcal{V}^{j} \to \mathcal{V}^{\ell}$. Let us decompose UIR's j and ℓ of M into irreducible representations $s \in \hat{\mathcal{U}}$ of $\mathcal{U}$,

$$\mathcal{V}^{\ell} = \sum_{\substack{s \in \hat{\mathcal{U}} \\ s \subset \ell}} \mathcal{U}^{s} \quad ; \quad \mathcal{V}^{j} = \sum_{\substack{s \in \hat{\mathcal{U}} \\ s \subset j}} \mathcal{U}^{s} \tag{10.15}$$

Consider the projection operators $\pi(\ell s)$ and their adjoints, viz. imbeddings $\pi^{*}(\ell s)$,

$$\pi(\ell s) : \mathcal{V}^{\ell} \longrightarrow \mathcal{U}^{s} \quad ; \quad \pi(\ell s)^{*} : \mathcal{U}^{s} \longrightarrow \mathcal{V}^{\ell} . \tag{10.16a}$$

Every $\mathcal{U}$-invariant map from V^j to V^ℓ is a linear combination of matrices

$$\hat{t}^{js} = \pi^*(\ell s)\pi(js) \quad ; s \in \hat{\mathcal{U}} \; ; \; s \subset \ell , s \subset j , \qquad (10.16b)$$

This solves the classification problem since the Clebsch-Gordan maps V^{js} are determined by $\hat{t}^{js}$. The labels j, s run through $j \in \hat{M}$, $s \in \hat{\mathcal{U}}$ and such that $j \subset \ell_1 \otimes \tilde{\ell}_2$, $s \subset j$ and $s \subset \ell$. The symbol $\subset$ is to be read as "is contained in...".

It remains to derive explicit formulae. We apply Eq. (10.14a) to express the kernels $t^{js}_{\#}$ in terms of $\hat{t}^{js}$. Given $g \in G'$ we have to find h in H and p in P such that $g = h\tilde{n}_{-\hat{x}}p^{-1}$. Every g in G' may be written in the form $g = \tilde{n}nma$ with $n \in N$ etc. The factor ma is readily absorbed into p^{-1}. It suffices therefore to consider $g = \tilde{n}_x^{-1} n_{wy}^{-1}$.

We seek h in H, p in P such that

$$h\tilde{n}_{-\hat{x}}p^{-1} = \tilde{n}_x^{-1} n_{wy}^{-1} . \qquad (10.17a)$$

The l.h.s. is $\tilde{n}^{-1}_{h\hat{x}} h p^{-1}$. By uniqueness of the Bruhat decomposition we must have $h\hat{x} = x$ and $hp^{-1} = n_{wy}^{-1}$. The solution of (10.17a) is thus

$$(h,p) = (h(x), n_{wy}h(x)) \text{ with } h(x) \in H \text{ such that } h(x)\hat{x} = x . \qquad (10.17b)$$

With that $t^{js}_{\#}(\tilde{n}_x^{-1} n_{wy}^{-1}) = t^{js}_{\#}(h\tilde{n}_{\hat{x}}^{-1}p^{-1}) = D^\chi(p)\hat{t}^{js}L^j(h)^{-1}$ whence

$$t^{js}_{\#}(\tilde{n}_x^{-1} n_{wy}) = D^\chi(h(x))\hat{t}^{js}L^j(h(x))^{-1} \equiv \hat{t}^{js}(x) \qquad (10.18)$$

independent of y. By definition (10.18),

$$\hat{t}^{js}(amx) = |a|^{-h-c-\sigma} D^\ell(m)\hat{t}^{js}D^j(m)^{-1} \; ; \; \hat{t}^{js}(\hat{x}) \equiv \hat{t}^{js} \qquad (10.18')$$

Let us now go back to Eq. (10.6a) for t^{js}. We insert (10.8) and make a change in the variable of integration.

This gives

$$(t^{js}F)(g_1) = \int dg_2\, t_{\#}^{js}(g_2^{-1})\, \tilde F(g_1 g_2).$$

Corresponding to the Bruhat decomposition $g_2 = n\tilde n m a$ the Haar measure factorizes as we know, $dg_2 = dm\,da\,dn\,d\tilde n$. So $(t^{js}F)(g_1) =$

$$\int dn\,d\tilde n\,dm\,da\; t_{\#}^{js}(a^{-1}m^{-1}\tilde n^{-1}n^{-1})\,\tilde F(g_1 n\tilde n m a) = \iint dn\,d\tilde n\; t_{\#}^{js}(\tilde n^{-1}n^{-1})\,L^{j}(ma)\,\tilde F(g_1 n\tilde n m a)\,dm\,da$$

by covariance (10.10). The integration over MA can now be performed with (10.5), whence

$$(t^{js}F)(g_1) = \int_N dn \int_{\tilde N} d\tilde n\; t_{\#}^{js}(\tilde n^{-1}n^{-1})\, F(g_1 n\tilde n) \tag{10.19}$$

We write $n = n_{wv}$, $\tilde n = \tilde n_x$, $dn = (v^2)^{-2h} dv$, $d\tilde n = dx$ and insert expression (10.18) for $t_{\#}$ to obtain

$$(t^{js}F)(n_{wy}\tilde n_z) = \int (v^2)^{-2h} dv\,dx\; \hat t^{js}(x)\, F(n_{wy}\tilde n_z n_{wv}\tilde n_x).$$

We simplify the argument of F by using identity (9. 8) to reexpress $\tilde n_z n_{wv}$. This gives $n_{wy}\tilde n_z n_{wv}\tilde n_x =$ $= n_{wy} n_{w(z+v)} \tilde n_{z'} h \tilde n_x = n_{wy+w(z+v)} \tilde n_{z'} \tilde n_{hx} h$ with z', h as indicated below. H -covariance (9.13) of F gives then finally

$$(t^{js}F)(n_{wy}\tilde n_z) = \int (v^2)^{-2h} dv\,dx\; \hat t^{js}(x)\, L^{j}(h)^{-1} F(n_{wy+w(z+v)}\tilde n_{z'+hx})$$

$$\text{with}\quad z' = w(wz - w(z+v)),\quad h = h_{v+z}\,h_{wv}. \tag{10.20a}$$

This formula will be saved for later use. In the special case $wy = z = 0$ we obtain instead

$$(t^{js}F)(e) = \int (v^2)^{-2h} dv\,dx\; \hat t^{js}(x)\, F(n_{wv}\tilde n_x). \tag{10.20b}$$

Eq. (10.20b) supplies us also with a formula for V^{js} .
Let $F = Q_j f$ for f in $\mathcal{C}_{\chi_1}\otimes\mathcal{C}_{\chi_2}$. Using the explicit formula (9. 9c) for $Q_j f = C(l_1 \tilde l_2; j)\,Qf$ we obtain for $V^{js}f = t^{js}Q_j f$

$$(V^{js}f)(e) = \int (y'^2)^{-2h} dy'\,dx'\; \hat t^{js}(x')\, C(l_1\tilde l_2; j)\,[D^{\chi_1}(h)\otimes D^{\chi_2}(h_{wy'})]\, f(x_1 x_2) \tag{10.21}$$

with $h \in H$, x_1, x_2 functions of x', y' as given in (9. 9c).
We reinsert the definition (10.18) of $\hat{t}^{js}(x)$ and use the M-invariance (9.12a) of the Clebsch-Gordan map $C(\ell_1\tilde{\ell}_2;j)$ to write

$$\hat{t}^{js}(x)C(\ell_1\tilde{\ell}_2;j)[D^{\chi_1}(h)\otimes D^{\chi_2}(h_{wy'})] = D^{\chi}(h(x'))\,\hat{t}^{js}C(\ell_1\tilde{\ell}_2;j)[D^{\chi_1}(h(x')^{-1}h)\otimes D^{\chi_2}(\tilde{h}(x')^{-1}h_{wy'})]$$

with $\tilde{h}(x') \equiv wh(x')w^{-1}$, h as before. (10.21a)

We split $h(x') = a(x')m(x')$ with $a(x') \in A$ and $m(x') \in M$ .Then

$|a(x')| = |x'|$ and $m(x')\hat{x} = x'/|x'|$ for our standard $\hat{x} = (\underset{\sim}{0},1)$.
Similarly $h_{wy'} = a_{wy'}m_{wy'}$; $h = am$. By the definition (9.19c) of h

$a = a_{y'}a_{wy} = a_{x'}a_{wx_1}$; $m = m_{x'}m_{wx_1}$. By definition, $D^{\chi}(ma) = |a|^{-h-c}D^{\ell}(m)$ etc.

We insert this in (10.21a). We have $|a(x')| = |x'| = |wx_1 - wx_2|^{-1} = |x_1||x_2||x_1-x_2|^{-1}$

and $|a| = y'^2|y|^{-2} = |x_2|^2|x_1-x_2|^{-2}$, and $|a_{wy'}| = |x_2|^{-2}$. So

$$\text{r.h.s. of (10.21a)} = |x_1|^{-h+c_1-c_2-c}|x_2|^{-h-c_1+c_2-c}|x_1-x_2|^{3h+c_1+c_2+c} \cdot \qquad (10.21b)$$
$$\cdot D^{\ell}(m(x'))\,\hat{t}^{js}C(\ell_1\tilde{\ell}_2;j)[D^{\ell_1}(m(x')^{-1}m)\otimes D^{\ell_2}(wm(x')^{-1}wm_{wx_2})]$$

with $x' = w(wx_1 - wx_2)$ as before, and $m = m_{x'}m_{wx}$.

We abbreviate

$$\check{t}^{js} \equiv \hat{t}^{js}C(\ell_1\tilde{\ell}_2;j) = \pi^*(\ell s)\pi(js)C(\ell_1\tilde{\ell}_2;j), \qquad (10.21c)$$

Matrices $\check{t}^{js}$ are $\mathcal{U}$-invariant maps $V^{\tilde{\ell}_1}\otimes V^{\tilde{\ell}_2} \mapsto V^{\ell}$. This follows from the $\mathcal{U}$-invariance of $\hat{t}^{js}$ and (9.22a) upon noting that $D^{\ell_1}(u) \equiv D^{\ell_1}(\theta u\theta) = D^{\tilde{\ell}_1}(u)$ since elements of $\mathcal{U}$ ("rotations") commute with the "time-reflection" θ .

We introduce an M-covariant version $\check{t}^{js}(x)$ by

$$\check{t}^{js}(m\hat{x}) = D^{\ell}(m)\,\check{t}^{js}\,[D^{\tilde{\ell}_1}(m)^{-1}\otimes D^{\tilde{\ell}_2}(m)^{-1}] \qquad (10.21d)$$

and the additional stipulation that $\check{t}^{js}$ is to be independent of the length of x, i.e. depends only on $x/|x|$. $\check{t}^{js}(x)$ is invariant under the little group $\mathcal{U}_x$ of x in M in the sense that

$$\mathcal{D}^{\ell}(u)\,\check{t}^{js}(x)[\mathcal{D}^{\tilde{\ell}_1}(u)\otimes\mathcal{D}^{\tilde{\ell}_2}(u)] = \check{t}^{js}(x) \qquad \text{for } u \text{ in } \mathcal{U}_x .$$

Let us abbreviate $wm(x')w^{-1} = \tilde{m}(x')$. We note the identity

$$m(x')^{-1}m = \tilde{m}(x')^{-1}m_{wx_1} \qquad \text{for } m = m_x, m_{wx_1} .$$

Indeed, for m in M and x arbitrary, $mm_x = m_{mx}\tilde{m}$ with $\tilde{m} = wmw^{-1}$ by (10.27a).

So $m(x')^{-1}m_{x'}m_{wx_1} = m_{\hat{x}}\,\tilde{m}(x')^{-1}m_{wx_1} = \tilde{m}(x')^{-1}m_{wx_1}$ since $m(x')\hat{x} = x'$ and $m_{\hat{x}} = \mathbb{1}$.

We insert this identity and (10.21d) into (10.21b) to obtain

$$\text{r.h.s. of (10.21a)} = |x_1|^{-h+c_1-c_2-c}\,|x_2|^{-h-c_1+c_2-c}\,|x_1-x_2|^{3h+c_1+c_2+c}$$
$$\check{t}^{js}(w(wx_1-wx_2))\,[\mathcal{D}^{\ell_1}(m_{wx_1})\otimes\mathcal{D}^{\ell_2}(m_{wx_2})] .$$

This can be inserted into (10.21) to give the final result. We define the CG-kernel (10.22) by

$$(V^{js}f)(\tilde{n}_x) = \int dx_1dx_2\, V^{js}(x_1\tilde{\chi}_1 x_2\tilde{\chi}_2 ; x\chi)\, f(x_1x_2). \qquad (10.22)$$

The Clebsch Gordan kernel for $\chi_1 = [\ell_1 c_1]$ $\chi_2 = [\ell_2 c_2]$, $\chi = [\ell c]$ is then given by

$$V^{js}(x_1\tilde{\chi}_1 x_2\tilde{\chi}_2 ; 0\chi) = |x_1|^{-h+c_1-c_2-c}\,|x_2|^{-h-c_1+c_2-c}\,|x_1-x_2|^{-h+c_1+c_2+c}$$
$$\cdot\,\check{t}^{js}\left(\frac{x_1}{x_1^2}-\frac{x_2}{x_2^2}\right)[\mathcal{D}^{\ell_1}(m_{wx_1})\otimes\mathcal{D}^{\ell_2}(m_{wx_2})] \qquad (10.23)$$

with $m_{wx} = -\Theta r(x) \in M$, $r(x)^{\mu}{}_{\nu} = -\delta^{\mu}{}_{\nu} + 2x^{\mu}x_{\nu}/|x|^2$.

We have made use of the fact that $\check{t}^{js}(x)$ is independent of the length of its argument by definition. The kernel (10.2) is translation invariant, so it suffices to know it for $x=0$. Summing up we have

Theorem 10.1. Let $\mathcal{U} \approx$ Spin $(2h-1) \subset M$ the centralizer in $M \approx$ Spin (2h) of reflections θ. Denote the set of their UIR's by $\hat{\mathcal{U}}$, $\hat{M}$ as usual.
The Clebsch Gordan kernels V^{js} for the Kronecker product of two elementary representations $V^{js}: [l_1 c_1] \otimes [l_2 c_2] \longrightarrow [l c]$ of $G \approx$ Spin (2h + 1,1) are labelled by j, s,

$$j \in \hat{M},\ s \in \hat{\mathcal{U}} \quad \text{and such that} \quad j \in l_1 \otimes \tilde{l}_2,\ s \subset j \text{ and } s \subset l,$$

with $\tilde{l}_2$ the mirror image of $l_2 \in \hat{M}$. The kernels are given explicitly by Eq. (10.23) above. The M-covariant projection-imbedding-operator $\check{t}^{js}(x)$ is independent of the length of its argument by definition and is given in terms of Clebsch Gordan coefficient (9.12a) for M and projection operators on UIR's of $\mathcal{U}$ (cp. (10.16a,b)) by Eqs. (10.21c,d), with $\tilde{D}^{l}(m) \equiv D^{l}(\theta m \theta)$, D^{l} the representation matrix for UIR l of M acting in vector space $V^{l} \equiv V^{\tilde{l}}$.

As an example, consider the special case $l_1 = l_2 = 0$ (trivial representation). Then $l_1 \otimes \tilde{l}_2 = 0$, so $j=0$ and $s=0$, and l must be a completely symmetric tensor representation in order that $s \subset l$. Vector spaces V^{l_1} and V^{l_2} are 1-dimensional. Linear maps $V^{l_1} \otimes V^{l_2} \longrightarrow V^{l}$ may thus be identified with vectors in V^{l}, so CG-kernels $V(\ldots)$ take values in V^{l}, and $\check{t}^{00}$ is a normalized $\mathcal{U}$-invariant vector $\check{t}^{00} = v_0 \in V^{l}$.

Completely symmetric tensor representations l of M have a unique (normalized) $\mathcal{U}$-invariant vector v_0. The components of $y^{l}(m\hat{x}) = D^{l}(m) v_0$ are called (zonal) spherical functions.

Corrollary 10.2. In the special case $l_1 = l_2 = 0$ (trivial representation) a nonvanishing CG-kernel exists only for $\chi = [l, c]$, where l is a completely symmetric tensor representation of M.

It is unique up to normalization (having $j=s=0$), viz.

$$V(x_1\chi_1\, x_2\chi_2\,;0\chi) = |x_1|^{-h-c+c_1-c_2}\,|x_2|^{-h-c-c_1+c_2}\,|x_1-x_2|^{-h+c+c_1+c_2}\cdot \eta^\ell(\lambda)$$

for $\chi_1=[0c_1]$, $\chi_2=[0c_2]$, with $\hat{\lambda} = \frac{|x_1||x_2|}{|x_1-x_2|}\left(\frac{x_1}{x_1^2}-\frac{x_2}{x_2^2}\right)$.

For completely symmetric tensor representations ℓ, vectors in V^ℓ may be considered as homogeneous polynomials in a complex isotropic ($\zeta^2=0$) $2h$-vector ζ, cp. Sec. 2A and Appendix A. The spherical functions are

$$\eta^\ell(\lambda) = c_\ell(\lambda\zeta)^\ell \quad ; \quad c_\ell^2 = \frac{4^\ell (h-1)_\ell\,(h)_\ell}{\ell!\,(2h-2)_\ell}\,.$$

So the CG-kernel of Corollary 10.2 becomes in this language

$$V(x_1\chi_1\, x_2\chi_2\,;0\chi) = c_\ell\,|x_1|^{-h-c+c_1-c_2+\ell}\,|x_2|^{-h-c-c_1+c_2+\ell}\,|x_1-x_2|^{-h+c+c_1+c_2-\ell}\left\{\zeta\cdot\left(\frac{x_1}{x_1^2}-\frac{x_2}{x_2^2}\right)\right\}^\ell. \tag{10.24}$$

10B. Application of the Plancherel theorem to the Kronecker product of two principal series representations.

In this section we consider the tensor product of two principal series representations of G. The expansion formula for this case will be deduced from the Plancherel theorem for the (left) regular representation, cp. Sec. 8.
We will restrict our attention to even number of spacetime dimensions 2h for now.

In Sec. 9 we showed that a Kronecker product of two principal series representations $\chi_1 \otimes \chi_2$ may be considered as a unitarily induced representation on G/MA. Because the inducing representation may be decomposed into irreducibles, it suffices to consider representations $\Xi^{j\sigma}$ which are induced by unitary irreducible representations $L^j(ma) = |a|^{-\sigma} D^j(m)$ of MA with $\sigma = c_1 - c_2$.

We show that every such representation is contained in the regular representation of G in the sense that the regular representation may be decomposed in a direct integral of such induced representations, and the integral runs over all of them, with multiplicity $d_j = \dim V^j > 0$.

The regular representation consists of complex valued square integrable functions f on G. Let $j \in \hat{M}$ be a UIR of M in V^j and v^j_α be an orthonormal basis in V^j. We consider the Fourier-Mellin transform of f for imaginary σ

$$f^{j\sigma}_\alpha(g) = \int_{MA} dm\,da\; |a|^{-\sigma} D^j(m) v^j_\alpha\, f(gma) \equiv (E^{j\sigma}_\alpha f)(g). \tag{10.25}$$

Functions $f^{j\sigma}_\alpha$ take values in V^j and have covariance property

$$f^{j\sigma}_\alpha(gma) = |a|^{\sigma} D^j(m^{-1}) f^{j\sigma}_\alpha(g) = L^j(ma)^{-1} f^{j\sigma}_\alpha(g). \tag{10.26}$$

This shows that they transform according to induced representation on G/MA. Mellin-inversion formula and Peter-Weyl theorem (i.e. orthogonality and completeness of representation functions for MA) provide the completeness relation

$$\int dg\, |f(g)|^2 = \sum_{j\in \hat M} \frac{d_j}{2\pi i} \int_{-i\infty}^{i\infty} d\sigma \int_{G/MA} d\dot g \sum_\alpha \langle f_\alpha^{j\sigma}(g), f_\alpha^{j\sigma}(g)\rangle \; ; \tag{10.27}$$

$\langle\, ,\rangle$ is the scalar product on V^j and $d_j = \dim V^j$. The integrand on the r.h.s. depends on g only through its coset $\dot g = gMA$. If $dg = dn\,d\tilde n\,dm\,da$ in terms of the Bruhat factors $g = n\tilde n m a$ then $d\dot g = dn\,d\tilde n$.

We may conclude that almost every unitarily induced representation on G/MA decomposes into irreducibles which appear in the decomposition of the regular representation. For <u>even</u> dimension 2h this means that only principal series representations appear in the decomposition. Moreover we can apply the expansion formula for square integrable functions $f(g)$ on G which effects their decomposition into functions which transform irreducibly. Applying to it the integral operator $E_\alpha^{j\sigma}$ of (10.25) we will obtain an expansion formula for functions $f^{j\sigma}$ in $\Xi^{j\sigma}$ which is certainly true in a distribution theoretic sense, i.e. (roughly) after smearing over σ with a test function $\xi(\sigma)$ such that $\int d\sigma\, \xi(\sigma) \langle v^j_\beta, f^{j\sigma}(g)\rangle$ are square integrable. We shall take it for granted that it holds also pointwise in σ whenever it makes sense. (A rigorous argument is known for the case of complex semisimple groups [N1, WS].)

Let then F^j be in $\Xi^{j\sigma}$, $\sigma = c_1 - c_2$, and let $g_1 = n_{wy}\tilde n_x$.

Define f_j in $\Xi^{j\sigma}$ by $f_j(g) = F^j(g_1 g)$

whence $$f_j(e) = F^j(n_{wy}\tilde n_x). \tag{10.28}$$

F^j and f_j are functions on G with covariance property

$$F^j(gma) = L^j(ma)^{-1}F^j(g) = |a|^\sigma D^j(m^{-1})F^j(g)\text{; same for } f_j. \tag{10.29}$$

The expansion formula alluded to above reads according to Sec. 8A

$$f_j(e) = \frac{1}{2} \sum\!\!\!\!\!\!\int d\chi \, \Theta_\chi(f_j) \quad \text{with } \chi = [\ell, c] \;, \; \sum\!\!\!\!\!\!\int d\chi \equiv \sum_{\ell \in \hat{M}} \frac{1}{2\pi i} \int_{-i\infty}^{+i\infty} \rho(\ell, c)\, dc \tag{10.30a}$$

where ρ is the Plancherel weight given by Theorem 8.1.
Θ_χ is the character of the principal series representation χ ; it is given by (3.18) and (3.9).

$$\Theta_\chi(f) = \int_K dk \, \mathrm{tr}\, F_f^\chi(k,k) \qquad \text{with}$$

$$\mathrm{tr}\, F_f^\chi(k_1, k_2) = \pi^h \Gamma(h)\Gamma(2h)^{-1} \int dm\, da\, dn\, |a|^{h-c} \xi^\ell(m) f(k_1 n m a k_2^{-1}) \tag{10.30b}$$

where $\xi^\ell(m) = \mathrm{tr}\, D^\ell(m)$

if Haar measure is normalized as $dg = dn\, d\tilde{n}\, dm\, da$ in terms of the Bruhat factors. K is the maximal compact subgroup of G .

It is convenient to translate formula (10.30b) to the noncompact picture (cp. Sec. 2D).

We write $k = k_z m$ with $m \in M$ and $k_z = \tilde{n}_z a(z)^{-1} n_z$ $|a(z)| = (1+z^2)^{-1}$. This is the Bruhat decomposition of $k \in K$ as we know, cp. (1.31) and (1.30c). Normalized Haar measure on K becomes $dk = \pi^{-1}\Gamma(2h)\Gamma(h)^{-1} |a(z)|^{2h} dz\, dm$
Inserting this in (10.30b) gives

$$\Theta_\chi(f) = \int |a(z)|^{2h} dz\, dm\, da\, dn\, |a|^{h-c} \xi^\ell(m) f(\tilde{n}_z a(z)^{-1} n_z^{-1} n\, m a n_z a(z) \tilde{n}_z^{-1}) \,.$$

We make a change of variables in the n -integration. Write first $n' = n_z^{-1} n\, n_{\tilde{m} a^{-1} z}$ with $\tilde{m} = w m w^{-1}$ then $dn = dn'$. Put now $n'' = a(z)^{-1} n' a(z)$, then $dn'' = |a(z)|^{2h} dn'$. Since $m a n_z = n_{\tilde{m} a^{-1} z}\, m a$ this gives

$$\Theta_\chi(f) = \int_{\mathbb{R}^{2h}} dz \int_{MAN} dm\, da\, dn\, |a|^{h-c} \xi^\ell(m) f(\tilde{n}_z n m a \tilde{n}_z^{-1}) \,, \tag{10.31}$$

where $\xi^\ell(m) = \mathrm{tr}\, D^\ell(m)$ the character of $\ell \in \hat{M}$ as before. This is the desired formula for the principal series character in the noncompact picture.

We will apply this to the function f_j defined by (10.28).

This is a vector valued function, with values in $\mathcal{V}^j$. The above formulae apply to its individual components in an arbitrary basis. We shall use vector notation, $\Theta_\chi(f_j)$ is then also a vector in $\mathcal{V}^j$. Writing $n = n_{wv}$, $dn = (v^2)^{-2h} dv$, etc. we get

$$\Theta_\chi(f_j) = \iint_{\mathbb{R}^{4h}} (v^2)^{-2h} dv dz \int_{MA} dm da |a|^{h-c} \xi^\ell(m) F^j(n_{wy} \tilde{n}_{x+z} n_{wv} m a \tilde{n}_z^{-1}) . \tag{10.32}$$

We pose ourselves the problem of expressing this in terms of the Clebsch-Gordan kernels which were determined in the previous section*). This will be done by using covariance condition (10.29) to carry out the integration over a subgroup U_z of M.

Let $U_z \subset M$ be the little group of z , it consists of u in M such that $uz = z$, hence $u\tilde{n}_z^{-1} = \tilde{n}_z^{-1} u$. If we write $m = m_1 u$ then by (10.29)

$$F^j(n_{wy} \tilde{n}_{x+z} n_{wv} m a \tilde{n}_z^{-1}) = D^j(u)^{-1} F^j(n_{wy} \tilde{n}_{x+z} n_{wv} m_1 a \tilde{n}_z^{-1}) \tag{10.33a}$$

$$\text{for} \quad m = m_1 u \; , \; u \in U_z \; .$$

The coset space $M/U_z \approx S^{2h-1}$ is the unit sphere in 2h dimensions, with normalized measure $d\dot{m}$. An integral of the form

$$\int_M dm\, \varphi(m) = \int_{M/U_z} d\dot{m}\, \overline{\varphi(m)} \quad \text{defines} \quad \overline{\varphi(m)} = \int_{U_z} du\, \varphi(mu) \tag{10.33b}$$

and $\overline{\varphi(m)}$ depends on m only through the coset $\dot{m} = mU_z$.

We shall use the following identity

Lemma 10.3. Let $\check{z} = z/|z|$ and $\hat{t}^{js}(\check{z})$ be the U_z-invariant map from $\mathcal{V}^j$ to $\mathcal{V}^\ell$ defined by (10.16a,b) and (10.18'). The adjoint map (=hermitean conjugate matrix) from $\mathcal{V}^\ell$ to $\mathcal{V}^j$ is denoted by $\hat{t}^{js}(\check{z})^*$. With this notation

*) It is implicit in the following considerations that the Clebsch Gordan kernels could have been deduced from (10.32) saving part of the labor of Sec. 2C.

$$\int_{U_z} du\, \xi^\ell(mu)\, D^j(u^{-1}) = \sum_{s \subset j,\ell} d(s)^{-1}\, \hat{t}^{js}(\check{z})^* D^\ell(m)\, \hat{t}^{js}(\check{z}) \,.$$

Summation is over UIR's s of $U_z \approx \mathrm{Spin}\,(2h-1)$ which are contained both in the UIR's j and ℓ of M; $d(s)$ is the dimension of representation s .

The proof of this lemme is relegated to Appendix D , it is a straightforward application of standard orthogonality relations of representation functions of compact Lie groups.

We make use of (10.33b) to split the m - integration in (10.32). We insert (10.33a) and carry out the integration over U_z with the help of lemma 10.3. This gives

$$\Theta_\chi(f_j) = \iint (v^2)^{-2h} dv\, dz \int_{M/U_z} d\dot{m} \int_A da\, |a|^{h-c} \sum_{s \subset \ell, j} d(s)^{-1}\, \hat{t}^{js}\left(\tfrac{z}{|z|}\right)^* D^\ell(m)\, \hat{t}^{js}\left(\tfrac{z}{|z|}\right) \cdot$$
$$\cdot F^j(n_{wy}\, \tilde{n}_{x+z}\, n_{wv}\, ma\, \tilde{n}_z^{-1}) \,.$$

(s -summation as in Lemma 10.3). (10.34a)

The argument of the integral depends on m only through the coset $\dot{m} = mU_z$.

From here on until Eq. (10.36) there is some amount of straightforward computation to be done. We will determine the Bruhat-decomposition of the argument g of F^j , use the covariance (10.29) of F^j and make some changes of variables in the integrations. In this we shall use the fact that MA acts transitively on the pointed Euclidean space $\mathbb{R}^{2h} \setminus \{0\}$. Finally, by comparison with Eq. (10.20a) of Sec. 10A we will be able to express the result in terms of $t^{js} F^j$.

Let us got to work then. We want to find the Bruhat factors of

$g = n_{wy}\, \tilde{n}_{x+z}\, n_{wv}\, ma\, \tilde{n}_z^{-1}$ in the order $n\tilde{n}ma$; $n \in N$, etc.

First we consider $\tilde{n}_{x+z}\, n_{wv}$. Identity (9.8) gives

$\tilde{n}_{x+z}\, n_{wv} = n_{w(x+z+v)}\, \tilde{n}_{z'}\, h_{x+z+v}\, h_{wv}$ with $wz' = w(x+y) - w(x+z+v)$.

We insert this in g and switch the h -factors through $\tilde{n}_z^{-1}$.
This gives

$g = n_{w(x+z+v)+wy}\, \tilde{n}'_{z'-h_{x+z+v}h_{wv}x'}\, h_{x+z+v}\, h_{wv}\, ma$, z' as above, x'=maz.

We insert this in (10.34a) and make use of covariance (10.29) of F^j .

$$\Theta_\chi(f_j) = \iint (v^2)^{-2h} dv\,dz \int d\dot{m} \int da\,|a|^{h-c} \sum_{s\subset j,\ell} d(s)^{-1} \hat{t}^{js}(\check{z})^* \mathcal{D}^\ell(m)\, \hat{t}^{js}(\check{z})$$

$$\cdot\, |a|^{-\sigma} \mathcal{D}^j(m)^{-1} L^j(h_{x+z+v} h_{wv})^{-1} F^j(n_{w(x+z+v)+wy}\, \tilde{n}_{z'} - h_{x+z+v} h_{wv} x')$$

$$\text{with } wz' = w(x+z) - w(x+z+v)\,,\quad x' = maz\,,\quad \check{z} = z/|z|\,. \tag{10.34b}$$

Next we use homogeneity and M-covariance (10.18') of $\hat{t}^{js}(z)$ to scale its argument and pull $\mathcal{D}^j(m)^{-1}$ through it:

$$\hat{t}^{js}(z/|z|)^* \mathcal{D}^\ell(m)\, \hat{t}^{js}(z/|z|)\, \mathcal{D}^j(m)^{-1} = |z|^{2h} |a|^{h+c-\sigma}\, \hat{t}^{js}(z)^*\, \hat{t}^{js}(maz)\,. \tag{10.35}$$

We also change variables of integration: The argument depends on m and a only through $maz = x'$, apart from a remaining explicit factor $|a|^{2h}$. MA acts transitively on pointed x-space $\mathbb{R}^{2h} \setminus \{0\}$, so integration over M/U_z and A may be replaced by integration over x'. $M/U_z \approx S^{2h-1}$, the sphere, so $d\dot{m} = d\Omega_{x'}$ is the angular integration (normalized to 1). Also $da = |a|^{-1} d|a| = |x'|^{-1} d|x'|$, whence $d\dot{m}\,da = |x'|^{-1} d\Omega_{x'} d|x'| = |x'|^{-2h} dx'$ and integration over x' runs over all of $\mathbb{R}^{2h}$ (the missing point 0 has measure 0).

Since $|a|^{2h}|z|^{2h} = |maz|^{2h} = |x'|^{2h}$ we get

$$\Theta_\chi(f_j) = \int (v^2)^{-2h} dv\,dz\,dx' \sum_{s\subset\ell,j} d(s)^{-1}\, \hat{t}^{js}(z)^*\, \hat{t}^{js}(x')\, L^j(h_v h^{-1}_{x+z+v})\,\cdot$$

$$F^j(n_{w(x+z+v)+wy}\, \tilde{n}_{z'} - h_{x+z+v} h_{wv} x')\,.$$

$$\text{with } wz' = w(x+z) - w(x+z+v)$$

By comparison with (10.20a) we see that $\Theta_\chi(f_j)$ may be expressed in terms of $t^{js} F^j$ as

$$\Theta_\chi(f_j) = \sum_{s\subset\ell,j} d(s)^{-1} \int dz\, \hat{t}^{js}(z)^* (t^{js} F^j)(n_{wy}\, \tilde{n}_{x+z}) \tag{10.36}$$

We now go through some of the steps of Sec. 10A in backward direction, expressing $\hat{t}^{js}(z)$ through the kernel $t^{js}(g_1, g_2)$ by (10.18), (10.8) viz.

$$\hat{t}^{js}(z) = t^{js}_{\#}(\tilde{n}_z^{-1}) = t^{js}(n_{wy}\tilde{n}_x\,, n_{wy}\tilde{n}_{z+x})\,.$$

Inserting the hermitean conjugate of this in (10.36) and using a new variable of integration $z' = z+x$ we get

$$\Theta_\chi(f_j) = \sum_s d(s)^{-1} \int dz' t^{js}(n_{wy}\tilde{n}_x ; n_{wy}\tilde{n}_{z'})^* (t^{js} F^j)(n_{wy}\tilde{n}_{z'})$$

Finally we use conformal covariance (of $t^{js}(. , .)$ and $t^{js} F^j$ which transform contragredient to each other) to write

$$\Theta_\chi(f_j) = \sum_{s \subset \ell, j} d(s)^{-1} \int dz\, t^{js}(n_{wy}\tilde{n}_x , \tilde{n}_z)^* (t^{js} F^j)(\tilde{n}_z) \tag{10.37}$$

Checking this amounts to a direct verification of the conformal invariance of the scalar product for a unitary principal series representation in the noncompact picture, cp. Sec. 2.D. The result (10.37) combined with (10.30a) (and 10.28) is the sought for <u>expansion formula for</u> (functions in the representation space of) <u>the unitarily induced representation $\Xi^{j\sigma}$ on G/MA.</u>

$$F^j(g) = \frac{1}{2} \sum\!\!\!\!\!\!\int d\chi \sum_{s \subset \ell, j} d(s)^{-1} \int_{\mathbb{R}^{2h}} dz\, t^{js}(g, \tilde{n}_z)^* (t^{js} F^j)(\tilde{n}_z) \,. \tag{10.38a}$$

We have derived this for g of the form $g = n_{wy}\tilde{n}_x \in N\tilde{N}$. It extends to general g by writing down the Bruhat-decomposition of g and using MA-covariance of F^j and $t^{js}(. , .)$.

The map $t^{js} : \Xi^{js} \mapsto \chi$ is an isometric map between unitary representation spaces, so it has an adjoint $t^{js*} : \chi \mapsto \Xi^{js}$ which effects the G-invariant imbedding; its integral kernel is just $t^{js}(\cdot,\cdot)^*$. Expansion formula (10.38a) may then be written for short

$$F^j = \frac{1}{2} \sum\!\!\!\!\!\!\int d\chi \sum_{s \subset \ell, j} d(s)^{-1}\, t^{js*} t^{js} F^j \quad ; \quad (\chi = [\ell, c]) \tag{10.38b}$$

In our last step we use the connection, established in Sec. 9, of the unitarily induced representations $\Xi^{j\sigma}$ and the Kronecker product $\chi_1 \otimes \chi_2$ of two unitary principal series representations. The connection is effected by operators Q_j , cp. end of Sec. 9 : Let f in $\chi_1 \otimes \chi_2$ and $F^j = Q_j f$. Then $f = \Sigma_j Q_j^* F^j$. Furthermore the CG-map $V^{js} = t^{js} \circ Q_j$, and Q_j^* is the adjoint of Q_j . Expansion formula (10.39b) gives then

$$f = \sum_j Q_j^* F^j = \frac{1}{2} \oint d\chi \sum_{j,s}\sum d(s)^{-1} Q_j^* t^{js*} t^{js} Q_j f$$

That is

$$f = \frac{1}{2} \oint d\chi \sum_{j,s}\sum d(s)^{-1} V^{js*} V^{js} f \tag{10.39}$$

where the sum is over UIR's j of M and s of U such that $j \subset \ell_1 \otimes \tilde{\ell}_2$ and $s \subset \ell, s \subset j$. Writing this out we have our final result

Theorem 10.4. Consider functions $f(x_1 x_2)$ of two Euclidean variables, contained in the Kronecker product $\chi_1 \otimes \chi_2$ of two unitary principal series representations of $G \simeq$ Spin (2h+1,1), h integer. [That is, $\chi_1 = [\ell_1, c_1]$, $\chi_2 = [\ell_2, c_2]$ with c_1, c_2 pure imaginary; f takes values in $\mathcal{V}^{\ell_1} \otimes \mathcal{V}^{\ell_2}$, $\mathcal{V}^{\ell}$ the vector space in which acts UIR ℓ of $M \simeq$ Spin (2h); f is square integrable in the sense of (9.3) and transforms as in (9.2b)]. Then f admits a conformal partial wave expansion in terms of principal series representations $\chi = [\ell, c]$ as follows. Let $V^{js}(\cdots)$ be the CG-kernels of Theorem 10.1. Then

$$\varphi_\chi^{js}(x) = \int_{\mathbb{R}^{4h}} dx_1 dx_2 \, V^{js}(x_1\tilde{\chi}_1\, x_2\tilde{\chi}_2 ; x\chi) f(x_1 x_2)$$

takes values in $\mathcal{V}^{\ell}$, and f admits expansion

$$f(x_1 x_2) = \frac{1}{2} \oint d\chi \sum_{j,s}\sum d(s)^{-1} \int dx \, V^{js}(x_1\tilde{\chi}_1\, x_2\tilde{\chi}_2 ; x\chi)^* \varphi_\chi^{js}(x) \quad , \quad (\chi = [\ell, c]).$$

Summation is over UIR's j of M which are contained in the Kronecker product $\ell_1 \otimes \tilde{\ell}_2$, $\tilde{\ell}_2$ the mirror image of $\ell_2 \in \hat{M}$, and over UIR's s of $U \approx$ Spin (2h-1) which are contained both in UIR's j and ℓ of M; $d(s)$ is the dimension of the UIR. s.

$$\oint d\chi = \sum_{\ell \in \hat{M}} (2\pi i)^{-1} \int_{-i\infty}^{i\infty} \rho(\ell, c)\, dc$$

with ρ the Plancherel weight of Theorem 8.1 the star $*$ denotes hermitean conjugation of a matrix (= adjoint map $\mathcal{V}^{\ell} \to \mathcal{V}^{\ell_1} \otimes \mathcal{V}^{\ell_2}$).

10.C Odd space time dimension 2h

In our discussion in Sec. 10.B we assumed that the number of space time dimensions 2h was even, viz. h = 1, 2, ... This implied, by the result of Sec. 8.A that only the principal series of unitary representations entered into the decomposition of the regular representation of $G \approx$ Spin (2h, 1) or SO (2h, 1). The same holds then for the Kronecker product of two unitary principal series representations.

For odd number of space time dimensions, the situation is different in general. According to the results reviewed in Sec. 8.B, the regular representation, and therefore the Kronecker products of principal series representations also, decompose in general into principal series and discrete series contributions. The contribution of the principal series is of the same form as before, only the explicit form of the Plancherel weight $\rho(\ell,c)$ changes (it has poles now). So it remains to discuss the possible contributions form the discrete series.

We consider the special case that $\chi_1 = [\ell_1, c_1]$ and $\chi_2 = [\ell_2, c_2]$ are class I representations, viz. $\ell_1 = \ell_2 = 0$ the trivial representation of M. They are of course representations of SO(2h+1,1). We show that in this case, discrete series representations do not appear.

The proof makes use of the following known fact reviewed in Sec. 7.

1) All elementary representations of G are simply reducible with respect to the maximal compact subgroup K = SO(2h+1). (In other words, each irreducible representation of K enters at most once in a given elementary representation χ of G - cf. Corollary 2.1).

2) Any discrete series representation U of G is inequivalent to its mirror image (i.e. to the representation obtained by space reflection) (Theorem 7.3).

3) It follows from 1) and 2) that each irreducible representation of K which belongs to a discrete series representation U of G is inequivalent to its mirror image.

4) The Gel'fand-Zeitlin patterns for SO(n)(which provide a simple rule for the $M = SO(2h)$ content of $K = SO(2h+1)$) tell us that if an irreducible representation of K contains the trivial representation of M, then it is equivalent to its mirror image. Together with 3) this implies that a discrete series representation U of G never contains the trivial representation of M.

Finally we shall make use of the following

Reciprocity Theorem (see Warner [W3] vol. 1, Theorem 5. 3. 3.1).

Let G be a Lie group and H be a closed subgroup of G. Let L and U be differentiable representations of H and G, respectively. Let T_L be the representation of G(differentiably) induced by L. Define the intertwining number $i(T_1, T_2)$ of two representations T_1 and T_2 of G, acting in the Fréchet spaces S_1 and S_2, as the number of linearly independent invariant bilinear forms $B(f_1, f_2)$ on $S_1 \otimes S_2$. Then the intertwining number $i(T_L, U)$ is equal to $i(L, U|_H)$ ($U|_H$ being the restriction of the representation U to the subgroup H of G).

Assume now that the induced representation $T_L \approx \Xi^{j\sigma}$ on G/MA contains a discrete series representation U of G. Then there is a (projection) map of Hilbert spaces, $\Pi : \mathcal{H}^{j\sigma} \mapsto \mathcal{H}_U$, which commutes with the action of the group: $\Pi T_L(g) = U(g)\Pi$.

Using the scalar product on $\mathcal{H}_U$, we can define an invariant bilinear form $B(f, \varphi)$ on $\mathcal{H}^{j\sigma} \times \mathcal{H}_{\bar{U}}$ where $\bar{U}$ is the complex conjugate representation of U. Restricting the representation T_L and U to corresponding (dense) Fréchet subspaces $\mathcal{S}_L$ and S_U of $\mathcal{H}^{j\sigma}$ and $\mathcal{H}_U$ (in order to comply with the differentiability assumption of the Reciprocity Theorem) we end up with a (non-trivial) bilinear form B on $\mathcal{J}_L \otimes S_{\bar{\mathcal{U}}}$. We now apply the Reciprocity Theorem, choosing for H the inducing subgroup MA of G. We conclude that the number of non-trivial (linearly independent) invariant bilinear forms on $\mathcal{J}_L \otimes S_{\bar{\mathcal{U}}}$ is equal to the intertwining number $i(L, \mathcal{U}_{|MA})$, where L is the one-dimensional representation (9.14) of MA and $\mathcal{U}|_{MA}$ is the restriction to MA of the discrete series representation U of G. According to 4) above, this number is zero, since L is trivial on M. Thus, no invariant bilinear form $B(f, \varphi)$ exists, contrary to our assumption. This proves that $\Xi^{j\sigma}$ contains no discrete series representation.

The results of this section are summed up in the following

<u>Theorem</u> 10.5. The tensor product of two class I (zero spin) principal series representations of the (generalized) Lorentz group G is decomposable into principal series (unitary) representations of G only.

10.D. Analytic continuation in c_1 and c_2 .

In our applications to conformal partial wave expansions in QFT we will need the decomposition of the Kronecker-product $\chi_1 \otimes \chi_2$ of two elementary representations for two special cases

1) $\chi_1 = [0, c_1]$, $\chi_2 = [0, c_2]$ and in the complementary series (c_i real)
2) $\chi_1 = [0, c_1]$ and in complementary series, χ_2 in the unitary principal series.

We shall now derive expansion formulae for these cases from the result of Sec. 10.B by analytic continuation in c_1 and c_2. The procedure will be somewhat heuristic.

Consider Schwartz test functions of compact support $f(x_1 x_2)$ with values in $\mathcal{V}^{\ell_1} \otimes \mathcal{V}^{\ell_2} \simeq \mathcal{V}^{\ell_2}$. They may be considered as elements of the representation space for the Kronecker product $\chi_1 \otimes \chi_2 = [\ell_1, c_1] \otimes [\ell_2, c_2]$ for arbitrary c_1, c_2. They are distinguished by the fact that their asymptotic expansions as $x_1 \to \infty$ and/or $x_2 \to \infty$ vanish identically. Nevertheless they form a dense subset of the Hilbert space which carries the unitary representation in cases 1), 2) above. We consider the expansion of such functions first, and discuss extension to the whole Hilbert-space later on.

We shall use the notation

$$\chi^* = [\ell, -\bar{c}] \quad \text{if} \quad \chi = [\ell, c] \tag{10.40}$$

$\bar{c}$ is the complex conjugate of c . We note that $\chi = \chi^*$ if (and only if) χ belongs to the unitary principal series. Using this we can rewrite expansion formula Theorem 10.4 as follows ($\chi = [\ell, c]$)

$$f(x_1 x_2) = \frac{1}{2} \sum_{\ell} \frac{1}{2\pi i} \int_{-i\infty}^{i\infty} \rho(\ell, c)\, dc \sum_{j,s} d(s)^{-1} \int dx\, V^{js}(x_1 \tilde{\chi}_1^* x_2 \tilde{\chi}_2^* ; x\chi)^* \cdot \int dx_1' dx_2'\, V^{js}(x_1' \tilde{\chi}_1\, x_2' \tilde{\chi}_2 ; x\chi)\, f(x_1' x_2') \tag{10.41}$$

If $F(c)$ is a matrix valued analytic function of c then so is $F(-\bar{c})^*$.

We understand the integral $\int dx_1' dx_2' (\dots)$ in the distribution theoretic sense (to start with, it is defined by an ordinary convergent integral). The kernels $V^{js}(\dots)$ are distributions meromorphic in the complex parameters c, c_1, c_2, with poles as specified below. The integrand of the c-integral in (10.41) is therefore a meromorphic function of c, c_1, and c_2.
[One should discuss convergence properties of the integrals, but we will content ourselves with these remarks: The x-integration is well convergent at large x because it is conformal invariant and can be rewritten as a compact integral in the "compact picture" of Sec. 2C. The c-integration is also well convergent at large c because the partial waves of smooth functions fall off in $|c|$ quickly].
We can then analytically continue equation (10.41) in c_2 and c_1. It will continue to hold as it stands as long as no poles of the integrand cross the path of the c-integration. If such a crossing occurs, however, this will produce an extra contribution according to the residue theorem. (We are not interested in formulae where the path of the c-integration is deformed).

Lemma 10.6. For $\chi_1 = [\ell_1, c_1], \chi_2 = [\ell_2, c_2]$ the kernel $V^{js}(x_1\chi_1 x_2\chi_2 ; x\chi)$ is a distribution meromorphic in c, c_1, c_2 and without poles in the half plane $\mathrm{Re}(h + c_1 + c_2) > 0$ (10.42)

We omit the proof of this lemma; it follows the lines of ref. [M7]. Let us explore its consequences. It implies that no pole of the integrand in (10.41) crosses the path of the c-integration so long as we stay with c_1 and c_2 in the halfplane (10.42), and also with $-\bar{c}_1$ and $-\bar{c}_2$.

Proposition 10.7. The Kronecker product $\chi_1 \otimes \chi_2$ of two unitary elementary representations $\chi_1 = [\ell_1, c_1]$, $\chi_2 = [\ell_2, c_2]$ in the supplementary or principal series decomposes into principal series representations only for even number 2h of space time dimensions provided $|\mathrm{Re}\, c_1| + |\mathrm{Re}\, c_2| \leq h$.
The same is still true for odd dimension 2h if $\ell_1 = \ell_2 = 0$.

Our considerations so far show that this proposition is correct for the subspace of functions $f(x_1 x_2)$ as specified above. We extend the result first to all functions which belong to the representation space $\mathcal{C}_{\chi_1}$ resp. $\mathcal{C}_{\chi_2}$ of Sec. 2.D when considered as functions of the individual variables x_1 resp. x_2. This means that we give up the requirement that their asymptotic expansion at large x vanishes. In this way we obtain a space which is invariant under the action of G (i.e. a true representation space). To make the extension, one needs to observe that the x_1', x_2' -integrations will still converge at large x_1', x_2' because they are conformal invariant and can be rewritten as compact integrations in the compact picture of Sec. 2.C. The point at ∞ is no special point then any more.

Further extension to the whole Hilbert space is made by completion in the norm given by the appropriate scalar product. For the Kronecker product of two class I (scalar) complementary series representations it is given by

$$(f_1, f_2) = \int dx_1 dx_2 dx_1' dx_2' \, \bar{f}_1(x_1 x_2) G^{-1}_{\chi_1}(x_1 - x_1') G^{-1}_{\chi_2}(x_2 - x_2') f_2(x_1' x_2') \tag{10.43a}$$

and for the Kronecker product of a class I complementary series χ_0 with a principal series χ_2 one has

$$(f_1, f_2) = \int dx_1 dx_1' dx_2 < f_1(x_1 x_2), f(x_1' x_2) > G^{-1}_{\chi_0}(x_1 - x_1') \tag{10.43b}$$

where $< , >$ is the scalar product in $v^{l_1} \otimes v^{l_2} \simeq v^{l_2}$ here. The intertwining kernel $G^{-1}_{\chi} = G_{\tilde{\chi}}$ is given in Sec. 4.B , only the scalar case is needed here.

<u>Conclusion</u> 10.8. The expansion formula (10.41) holds true for functions with finite norm (f,f) given by (10.43a or b) if the hypothesis of proposition 10.7 is satisfied. Convergence is strong convergence in Hilbert space.

11. Special cases and further properties of the expansion formula

11.A The Clebsch-Gordan kernel for two class I representations. Symmetry and normalization.

In this section we shall start from special cases of the results of Sec. 10 and consider the Clebsch-Gordan decomposition of two class I ("scalar") representations

$$\chi_1 = [0, c_1] , \quad \chi_2 = [0, c_2] \tag{11.0}$$

of G (c_1 and c_2 satisfying $c_k^2 < h^2$, $k = 1, 2$, in other words each c_k is either pure imaginary or belongs to the interval $-h < c_k < h$, -- cf. Secs. 3.D., 5.C). For ease of future reference we shall first restate the results of Sec. 10. We shall however use different normalization conventions for the Clebsch Gordan kernels in order to exhibit symmetry properties of these kernels in the most convenient form.

Let $f(x_1, x_2) \in S_{K_0}$ (the space of infinitely smooth functions which vanish at coinciding arguments and are obtained by restriction to $\tilde{N} \times \tilde{N}$ of C^∞-function f on $G \times G$ having covariance property (9.1)). The Clebsch Gordan kernels $V(x_1\tilde{\chi}_1 x_2\tilde{\chi}_2 ; x\chi)$ allow to define vector valued functions

$$F_\chi(x) = \frac{1}{2} \iint V(x_1\tilde{\chi}_1 x_2\tilde{\chi}_2 ; x\chi) f(x_1, x_2)\, dx_1 dx_2 \tag{11.1}$$

such that

$$[T^\chi(g) F_\chi](x) = \frac{1}{2} \iint V(x_1\tilde{\chi}_1 x_2\tilde{\chi}_2 ; x\chi) [T(g) f](x_1, x_2)\, dx_1 dx_2 \tag{11.2}$$

where $T = T^{\chi_1} \otimes T^{\chi_2}$. The intertwining property (11.2) implies that V has to satisfy the covariance condition

$$[T^{\tilde{\chi}_1} \otimes T^{\tilde{\chi}_2} \otimes T^\chi (g) V](x_1\tilde{\chi}_1, x_2\tilde{\chi}_2 ; x\chi) = V(x_1\tilde{\chi}_1, x_2\tilde{\chi}_2 ; x\chi) \tag{11.3}$$

Such invariant 3-point functions have been constructed in the course of the study of conformal quantum field theory (see, e.g.[M10,T4]). In particular, it was shown [F1] , that if χ_1 and χ_2 are the zero spin representations (11.0), then a nonvanishing conformal invariant 3-point function only exists if χ is a type I (symmetric tensor) representation.

According to Corollary 10.2 and Eq. (10.24) the general conformal invariant 3-point function for the above class of representations can be written in the form

$$V(x_1\chi_1\, x_2\chi_2 ; x_3\chi\, \mathfrak{z}) = \frac{N_\ell(c_+,c_-,c)}{(2\pi)^h}\left(\frac{2}{x_{12}^2}\right)^{\frac{1}{2}(h-c+\ell)+c_+}\left(\frac{4}{x_{13}^2\, x_{23}^2}\right)^{\frac{1}{2}(h+c-\ell)}\left(\frac{x_{23}^2}{x_{13}^2}\right)^{c_-}(\lambda\cdot\mathfrak{z})^\ell \tag{11.4a}$$

$$= \frac{N_\ell(c_+,c_-,c)}{(\delta_\chi+c_-)_\ell(\delta_\chi-c_-)_\ell(2\pi)^h}\left(\frac{2}{x_{12}^2}\right)^{\frac{1}{2}(h-c+\ell)+c_+} D_\ell(\delta_\chi+c_-,\mathfrak{z}\cdot\nabla_1;\delta_\chi-c_-,\mathfrak{z}\cdot\nabla_2)\left[\left(\frac{2}{x_{13}^2}\right)^{\frac{1}{2}(h+c-\ell)+c_-}\left(\frac{2}{x_{23}^2}\right)^{\frac{1}{2}(h+c-\ell)-c_-}\right] \tag{11.4b}$$

where $\nabla_i = \frac{\partial}{\partial x_i}$, $i=1,2$, $c_\pm = \frac{1}{2}(c_1 \pm c_2)$, and D_ℓ is the polynomial

$$D_\ell(a,\lambda;b,\mu) = \sum_{\kappa=0}^{\ell}\binom{\ell}{\kappa}(a+\kappa)_{\ell-\kappa}(b+\ell-\kappa)_\kappa(-\lambda)^\kappa\mu^{\ell-\kappa}$$

$$= \ell!\,(\lambda+\mu)^\ell\, P_\ell^{(a-1,b-1)}\left(\frac{\mu-\lambda}{\mu+\lambda}\right) \quad ; \quad \delta_\chi = \frac{1}{2}(h+c-\ell). \tag{11.5}$$

[We have used again the notation (5.11); $P_\ell^{(\alpha,\beta)}$ is the Jacobi polynomial.]

Here χ_1 and χ_2 are the class I representations (11.0), $\mathfrak{z}$ belongs to the complex light cone $\mathbb{K}_{2h}$ (see (2.11)),

$$\lambda = \frac{\partial}{\partial x_3}\log\frac{x_{23}^2}{x_{13}^2} = 2\left(\frac{x_{13}}{x_{13}^2}-\frac{x_{23}}{x_{23}^2}\right), \quad x_{ik} = x_i - x_k . \tag{11.6}$$

The harmonic extension of $(\lambda\cdot\mathfrak{z})^\ell$ is given by (A.13) (n=2h)

$$H_\ell(\lambda,\zeta) = (\lambda\cdot\zeta)^\ell F\left(-\frac{\ell}{2},\frac{1-\ell}{2};2-h-\ell;\frac{\lambda^2\zeta^2}{(\lambda\cdot\zeta)^2}\right) = \frac{\ell!}{(h-1)_\ell}\left(\frac{\lambda^2\zeta^2}{4}\right)^{\frac{1}{2}\ell} C_\ell^{h-1}(\hat\lambda\cdot\hat\zeta),$$

$$\hat q = (q^2)^{-\frac{1}{2}} q \qquad (q = \lambda,\zeta).$$

We observe that Eqs. (11.1) (11.4) define a C^∞-function $F_\chi(x,\mathfrak{z})$ $(\in \mathcal{C}_\chi)$ provided that the exponents of x_{13}^{-2} and x_{23}^{-2} in V do not assume any of the positive (half) integer values

$$(h+c-\ell)\pm c_- = h+k, \quad k = 0, 1, 2, \ldots \tag{11.7}$$

The kernels V satisfy symmetry conditions which relate their values at equivalent points in the representation space. Indeed, it follows from the covariance properties of the V's and the intertwining operators G_χ (4.19) that (for example) the kernel

$$\int G_{\chi_1}(x_1-x_1')\, V(x_1'\tilde{\chi}_1, x_2\chi_2, x_3\chi)\, dx_1'$$

should be proportional to $V(x_1\chi_1, x_2\chi_2, x_3\chi)$ etc. We shall choose the normalization constant N_ℓ in (11.4) in such a way that the proportionality constant is equal to one. In other words we require

$$\int G_{\tilde{\chi}_1}(x_1-x_1')\, V(x_1'\chi_1, x_2\chi_2, x_3\chi)\, dx_1' = V(x_1\tilde{\chi}_1, x_2\chi_2, x_3\chi) \tag{11.8a}$$

and similar formulas for the arguments 2 and 3.

We shall further demand that N_ℓ is analytic in c_1, c_2 , and reduces to a phase factor for imaginary c_1, c_2, c , viz. principal series representations. Thus we demand that for imaginary c

$$\overline{N_\ell(-\bar{c}_+, -\bar{c}_-, c)}\, N_\ell(c_+, c_-, c) = 1 \tag{11.8b}$$

Such phase factors (and their analytic continuation) cancel out in the expansion formulae (Theorem 10.4 and Eq. (10.41)) of Sec. 10.) The constants c_ℓ appearing in (10.24) were omitted in (11.4). They will be absorbed in the Plancherel weight.

The expansion formula (10.41) and reality properties of kernels (11.4) imply then orthogonality relations of the following form

$$\iint dx_1 dx_2\, V(x_1\tilde{\chi}_1, x_2\tilde{\chi}_2, x\chi\mathfrak{z})\, V(x_1\chi_1, x_2\chi_2; x'\tilde{\chi}'\mathfrak{z}')$$

$$= \delta(\chi,\chi')\, \delta(x-x')\, (\mathfrak{z}\cdot\mathfrak{z}')^\ell + \delta(\chi,\tilde{\chi}')\, G^0_\chi(x-x'; \mathfrak{z},\mathfrak{z}') \; ; \tag{11.9}$$

here

$$\delta(\chi,\chi') = \delta_{\ell\ell'}\, \delta(\sigma-\sigma')\, \frac{2\pi}{\rho_\ell(i\sigma)} \quad , \quad \text{for} \quad \chi^{(\prime)} = [\ell^{(\prime)}, i\sigma^{(\prime)}] \, . \tag{11.10}$$

and $\rho_\ell(c)$ is the Plancherel weight:

$$\rho_\ell(c) = \frac{\Gamma(h+\ell)}{2(2\pi)^h \ell!} \left|\frac{\Gamma(h-1+c)}{\Gamma(c)}\right|^2 [(h+\ell-1)^2 - c^2] = \frac{\Gamma(h+\ell)}{2(2\pi)^h \ell!} n(\chi) n(\tilde{\chi}) \tag{11.11}$$

(for $c = i\sigma$).

The sum of two terms in the r.h.s. of (11.9) arises in the following way. Because of equivalence of principal series representations χ and $\tilde{\chi}$, every UIR appears twice in the partial wave expansion (10.41); because of amputation convention (11.8a) on leg 3, both contributions are equal. This fixes the relative size of the two terms on the r.h.s. of (11.9).

In deriving (11.9) from (10.41) one also needs the overall constant in the Plancherel weight (8.6). Conflicting answers to this are given in the literature. We have used the constant found by explicit calculation in [D.5].

Expansion formula (10.41) implies that the mapping given by (11.1) can be inverted with the result

$$f(x_1, x_2) = \sum\!\!\!\!\!\!\int d\chi \int dx \, V(x_1\chi_1, x_2\chi_2, x\tilde{\chi}) F_\chi(x) \tag{11.12}$$

where
$$\sum\!\!\!\!\!\!\int d\chi = \sum_{\ell=0}^{\infty} \int_{-i\infty}^{i\infty} \frac{dc}{2\pi i} \rho_\ell(\sigma) = \sum_{\ell=0}^{\infty} \int_{-\infty}^{\infty} \frac{d\sigma}{2\pi} \rho_\ell(\sigma) ; \quad \chi = [\ell, i\sigma] = [\ell, c] .$$

This is valid provided

$$|\mathrm{Re}\, c_1| + |\mathrm{Re}\, c_2| < h , \tag{11.13}$$

It is known for the ordinary Lorentz group that for $c_1 + c_2 > h$ the complementary series representation $c = c_1 + c_2 - h$ may contribute to the expansion [N1]. This suggests that at the boundary $c_1 + c_2 = h$ the expansion in principal series representations will still hold.

In the rest of this section we shall demonstrate that Eqs.(11.8) and (11.9) are compatible and fix $N_\ell(c_+, c_-, c)$ up to a sign.

We start with the exploitation of the symmetry property (11.8). The calculation is based on the integral formula

$$I(x_1\delta_1, x_2\delta_2, x_3\delta_3) = \frac{1}{(2\pi)^h}\int \frac{\Gamma(\delta_1)}{[\frac{1}{2}(x-x_1)^2]^{\delta_1}} \frac{\Gamma(\delta_2)}{[\frac{1}{2}(x-x_2)^2]^{\delta_2}} \frac{\Gamma(\delta_3)}{[\frac{1}{2}(x-x_3)^2]^{\delta_3}}\,dx =$$

$$= \int_0^\infty \frac{d\alpha_1}{\alpha_1}\int_0^\infty \frac{d\alpha_2}{\alpha_2}\int_0^\infty \frac{d\alpha_3}{\alpha_3}\, \frac{\alpha_1^{\delta_1}\alpha_2^{\delta_2}\alpha_3^{\delta_3}}{(k_1\alpha_1+k_2\alpha_2+k_3\alpha_3)^h}\exp\left\{-\frac{\alpha_1\alpha_2 x_{12}^2+\alpha_1\alpha_3 x_{13}^2+\alpha_2\alpha_3 x_{23}^2}{2(k_1\alpha_1+k_2\alpha_2+k_3\alpha_3)}\right\} =$$

$$(k_i \geq 0,\ \Sigma k_i > 0)$$

$$= \frac{\Gamma(h-\delta_1)\quad \Gamma(h-\delta_2)\quad \Gamma(h-\delta_3)}{(\frac{1}{2}x_{23}^2)^{h-\delta_1}(\frac{1}{2}x_{13}^2)^{h-\delta_2}(\frac{1}{2}x_{12}^2)^{h-\delta_3}} \quad \text{for } \delta_1+\delta_2+\delta_3 = 2h \tag{11.14}$$

(see [D2] [S8]) and on the identity

$$\frac{(\lambda\, \mathfrak{z})^\ell}{(\frac{1}{2}x_{13}^2)^\alpha(\frac{1}{2}x_{23}^2)^\beta} = \sum_{k=0}^{\ell}(-1)^k\binom{\ell}{k}\frac{\Gamma(\beta)}{\Gamma(\beta+k)}\left(\frac{2}{x_{13}^2}\right)^{\alpha+\ell-k}(\mathfrak{z}\cdot x_{13})^{\ell-k}(\mathfrak{z}\cdot\nabla_3)^k\left(\frac{2}{x_{23}^2}\right)^\beta =$$

$$= \sum_{k=0}^{\ell}(-1)^k\binom{\ell}{k}\frac{\Gamma(\alpha)\ \Gamma(\beta)}{\Gamma(\alpha+\ell-k)\Gamma(\beta+k)}\left(\frac{2}{x_{13}^2}\right)^\alpha(\mathfrak{z}\cdot\overleftarrow{\nabla}_3)^{\ell-k}(\mathfrak{z}\cdot\overrightarrow{\nabla}_3)^k\left(\frac{2}{x_{23}^2}\right)^\beta \tag{11.15}$$

(used for $\alpha = \frac{h+c-\ell}{2}+c_-$, $\beta = \frac{h+c-\ell}{2} - c_-$), where $\overleftarrow{\nabla}_3(\overrightarrow{\nabla}_3)$ differentiates with respect to x_3 to the left (right). Using the first equation (11.15) and (11.14) we find:

$$\int dx_2' \, V(x_1\chi_1, x_2'\chi_2; x_3\chi\mathfrak{z}) \, G_{\tilde{\chi}_2}(x_2-x_2') =$$

$$= \frac{N_\ell(c_+, c_-, c)}{(2\pi)^h} \sum_{k=0}^{\ell} (-1)^k \binom{\ell}{k} \frac{\Gamma\left(\frac{h+c-\ell}{2}-c_+\right)\Gamma\left(\frac{h-c+\ell}{2}+c_-\right)}{\Gamma\left(\frac{h-c+\ell}{2}+c_+\right)\Gamma\left(\frac{h+c-\ell}{2}-c_-+k\right)} \left(\frac{2}{x_{12}^2}\right)^{\frac{h-c+\ell}{2}+c_-} \times$$

$$\times \left(\frac{2}{x_{13}^2}\right)^{\frac{h+c-\ell}{2}+c_-} \left(\frac{2\mathfrak{z}x_{13}}{x_{13}^2}\right)^{\ell-k} (\mathfrak{z}\cdot\nabla_3)^k \left(\frac{2}{x_{13}^2}\right)^{c_2} \left(\frac{2}{x_{23}^2}\right)^{\frac{h+c-\ell}{2}-c_+} =$$

$$= \frac{N_\ell(c_+, c_-, c)\,\Gamma\left(\frac{h-c+\ell}{2}+c_-\right)}{(2\pi)^h\,\Gamma(c_2)\,\Gamma\left(\frac{h-c+\ell}{2}+c_+\right)} \left(\frac{2}{x_{12}^2}\right)^{\frac{h-c+\ell}{2}+c_-} \sum_{k=0}^{\ell} (-1)^k \binom{\ell}{k} \frac{1}{\Gamma\left(\frac{h+c-\ell}{2}-c_-+k\right)} \times$$

$$\times \sum_{j=0}^{k} \binom{k}{j} \frac{(\mathfrak{z}x_{13})^{\ell-j}(\mathfrak{z}x_{23})^j\,\Gamma(c_2+k-j)\,\Gamma\left(\frac{h+c-\ell}{2}-c_++j\right)}{\left(\frac{1}{2}x_{13}^2\right)^{\frac{h+c+\ell}{2}+c_--j}\left(\frac{1}{2}x_{23}^2\right)^{\frac{h+c-\ell}{2}-c_++j}} \,.$$

Changing the order of summation and using the sum rule

$$\sum_{k=j}^{\ell} (-1)^k \binom{\ell}{k}\binom{k}{j} \frac{\Gamma(\alpha+k)}{\Gamma(\beta+k)} = (-1)^j \binom{\ell}{j} \frac{\Gamma(\beta-\alpha+\ell-j)\,\Gamma(\alpha+j)}{\Gamma(\beta+\ell)\,\Gamma(\beta-\alpha)} \tag{11.16}$$

(see Appendix E) for $\alpha = c_2 - j$, $\beta = \frac{h+c-\ell}{2} - c_-$ we obtain

$$\int dx_2' \, V(x_1\chi_1, x_2', \chi_2, x_3\chi\mathfrak{z}) \, G_{\tilde{\chi}_2}(x_2-x_2')$$

$$= \frac{N_\ell(c_+, c_-, c)\,\Gamma\left(\frac{h+c+\ell}{2}-c_+\right)\Gamma\left(\frac{h-c+\ell}{2}+c_-\right)}{(2\pi)^h\,\Gamma\left(\frac{h-c+\ell}{2}+c_+\right)\Gamma\left(\frac{h+c+\ell}{2}-c_-\right)\left(\frac{1}{2}x_{12}^2\right)^{\frac{h-c+\ell}{2}+c_-}\left(\frac{1}{2}x_{13}^2\right)^{\frac{h+c-\ell}{2}+c_+}\left(\frac{1}{2}x_{23}^2\right)^{\frac{h+c-\ell}{2}-c_+}} =$$

$$= \frac{N_\ell(c_+, c_-, c)\,\Gamma\left(\frac{h+c+\ell}{2}-c_+\right)\Gamma\left(\frac{h-c+\ell}{2}+c_-\right)}{N_\ell(c_-, c_+, c)\,\Gamma\left(\frac{h-c+\ell}{2}+c_+\right)\Gamma\left(\frac{h+c+\ell}{2}-c_-\right)} V(x_1\,\chi_1, x_2\tilde{\chi}_2, x_3\chi\mathfrak{z}) \,. \tag{11.17}$$

Applying the obvious symmetry property

$$V(x_1\chi_1, x_2\chi_2, x_3\chi\mathfrak{z}) = (-1)^\ell \frac{N_\ell(c_+, c_-, c)}{N_\ell(c_+, -c_-, c)} V(x_2\,\chi_2, x_1\chi_1, x_3\,\chi\mathfrak{z}) \tag{11.18}$$

we can derive from (11.17) another relation of that type, involving integration over the first argument of V.

Combining these two equations and comparing with (11.8) we obtain

$$\frac{N_\ell(c_+,c_-,c)}{N_\ell(-c_+,-c_-,c)} = \frac{\Gamma\left(\frac{h-c+\ell}{2}+c_+\right)\Gamma\left(\frac{h+c+\ell}{2}+c_+\right)}{\Gamma\left(\frac{h-c+\ell}{2}-c_+\right)\Gamma\left(\frac{h+c+\ell}{2}-c_+\right)} \quad . \tag{11.19}$$

The consequences of the amputation identity (11.8a) on leg 3 can be derived in a similar way. It gives, for imaginary c

$$\frac{N_\ell(c_+,c_-,c)}{N_\ell(c_+,c_-,-c)} = \frac{\Gamma\left(\frac{h+c+\ell}{2}-c_-\right)\Gamma\left(\frac{h+c+\ell}{2}+c_-\right)}{\Gamma\left(\frac{h-c+\ell}{2}-c_-\right)\Gamma\left(\frac{h-c+\ell}{2}+c_-\right)} \tag{11.20}$$

We shall not reproduce the necessary computations here. (An independent derivation is contained in [D5]). In any case, (11.20) and the relative size of the two terms on the r.h.s. of (11.9) are merely a convention.

Eqs. (11.8b), (11.19) and (11.20) for the normalization factors are solved by

$$N_\ell(c_+,c_-,c) = \left[\frac{\Gamma\left(\frac{h-c+\ell}{2}+c_+\right)\Gamma\left(\frac{h+c+\ell}{2}+c_+\right)\Gamma\left(\frac{h+c+\ell}{2}-c_-\right)\Gamma\left(\frac{h+c+\ell}{2}+c_-\right)}{\Gamma\left(\frac{h-c+\ell}{2}-c_+\right)\Gamma\left(\frac{h+c+\ell}{2}-c_+\right)\Gamma\left(\frac{h-c+\ell}{2}-c_-\right)\Gamma\left(\frac{h-c+\ell}{2}+c_-\right)}\right]^{\frac{1}{2}} \tag{11.21}$$

The sign of the square root will be fixed in the following way. Let c_1 and c_2 be fixed real numbers such that

$$|c_1| + |c_2| < h \tag{11.22}$$

We regard N_ℓ as an analytic function in c with cuts along those intervals on the real c-axis for which the expression under the square root sign in (11.21) is non-positive. We demand further that for real c, for which

$$|c_1| + |c_2| + |c| < h+\ell \quad , \tag{11.23}$$

N_ℓ is positive. This rule determines N_ℓ completely in the cut c-plane. We note, however, that in most applications only various products of the N_ℓ's enter, which do not involve square roots.

11.B. Identities for the Clebsch-Gordan kernels at exceptional integer points

In the physical applications which we have in mind (see [M2, D4] and Chapter V) one needs analytic continuation of V not only in c_1 and c_2, but also in c to real values. In that case some differential identities between the V's at the exceptional integer points (3.3) play an important role. We shall describe below these identities excluding for the sake of simplicity the case $c_- = 0$, ν odd.

We shall first establish the relation

$$(3\,\nabla_3)^\nu V(x_1\chi_1, x_2\chi_2, x_3\bar{\chi}_{\ell\nu}3) = \text{sign}\left[\left(\tfrac{1-\nu}{2}+c_-\right)_\nu\right] V(x_1\chi_1, x_2\chi_2, x_3\chi'^{-}_{\ell\nu}3) \tag{11.24}$$

between the V's at exceptional points with $c < 0$.

To prove (11.24) we use the representation (11.4b) for V and remark that the operator $3\cdot\nabla_3$ is equivalent to $-(3\cdot\nabla_1 + 3\cdot\nabla_2)$ when applied to a translation invariant function of x_1, x_2, x_3 (which only depends on the coordinate differences). Then (11.24) is a consequence of the identities:

$$D_{\ell+\nu}\left(\tfrac{1-\nu}{2}-\ell+c_-, \lambda; \tfrac{1-\nu}{2}-\ell-c_-, \mu\right) = \left(\tfrac{1-\nu}{2}+c_-\right)_\nu (\lambda+\mu)^\nu D_\ell\left(\tfrac{1-\nu}{2}-\ell+c_-, \lambda; \tfrac{1-\nu}{2}-\ell-c_-, \mu\right), \tag{11.25a}$$

$$\frac{(\delta_\chi+c_-)_{\ell+\nu}(\delta_\chi-c_-)_{\ell+\nu}}{(\delta_\chi+c_-)_\ell(\delta_\chi-c_-)_\ell} = (-1)^\nu\left[\left(\tfrac{1-\nu}{2}+c_-\right)_\nu\right]^2 , \tag{11.25b}$$

where

$$\delta_\chi = \delta_{\ell\nu} = \frac{1-\nu}{2} - \ell \tag{11.26}$$

for both $\chi_{\ell\nu}^{-}$ and $\chi_{\ell\nu}^{\prime -}$ (where δ_χ is defined in (11.5)) and we use the relation

$$\ell!\, P_\ell^{(\delta_{\ell\nu}-1+c_-,\, \delta_{\ell\nu}-1-c_-)}\left(\frac{\mu-\lambda}{\lambda+\mu}\right) = \left(\frac{1-\nu}{2} - \ell + c_-\right)_\ell F\left(-\ell, -\ell-\nu; \frac{1-\nu}{2} - \ell + c_-; \frac{\lambda}{\lambda+\mu}\right)$$

(see Eq. 8.962.1 of ref. [G7]). Inserting (11.25) in (11.4b) we obtain

$$[-(\mathfrak{z}\nabla_1 + \mathfrak{z}\nabla_2)]^\nu V(x_1\chi_1, x_2\chi_2, x_3\chi_{\ell\nu}^{-}\mathfrak{z}) = \frac{N(\chi_{\ell\nu}^{-})}{N(\chi_{\ell\nu}^{\prime -})}\left(\frac{1-\nu}{2} + c_-\right)_\nu V(x_1\chi_1, x_2\chi_2, x_3\chi_{\ell\nu}^{\prime -}\mathfrak{z}) \tag{11.27}$$

where we use the short-hand notation $N(\chi)$ for the normalization constant (10.21). According to (10.21) that leads to (11.24).

Now we shall consider exceptional points with $c > 0$. For such points we shall derive the following relation:

$$\frac{\ell!}{(\ell+\nu)!}(\nabla_3 \partial_\zeta)^\nu V(x_1\chi_1, x_2\chi_2, x_3\chi_{\ell\nu}^{\prime +}\zeta) = \operatorname{sign}\left[\left(\frac{1-\nu}{2} + c_-\right)_\nu\right] \frac{(2h+\ell-2)_\nu}{(h+\ell-1)_\nu} V(x_1\chi_1, x_2\chi_2, x_3\chi_{\ell\nu}^{+}\zeta) \tag{11.28}$$

In order to prove (11.28) we shall use the x-space relation between the bilinear forms $\mathcal{B}_{\ell\nu}^{\prime +}$ and $\mathcal{B}_{\ell\nu}^{+}$ ((6.27) and (6.24)), which follows from (6.32)

$$(-1)^\nu \frac{\ell!^2}{(\ell+\nu)!^2}(\nabla \partial_\zeta)^\nu \mathcal{B}_{\ell\nu}^{\prime +}(x; \zeta, \partial_3)(\overleftarrow{\nabla}_3)^\nu = \frac{(2h+\ell-2)_\nu}{(h+\ell-1)_\nu} \mathcal{B}_{\ell\nu}^{+}(x; \zeta, \partial_3), \tag{11.29}$$

and the analytic continuation in c_1, c_2 and c of (11.8) (with $x_1\chi_1$ replaced by $x_3\chi$) to real c's for which χ_1 and χ_2 are complementary series representations and $\chi = \chi_{\ell\nu}^{(\prime)\pm}$:

$$\frac{1}{\ell!}\int \mathcal{B}_{\ell\nu}^{+}(x_{33'}; \zeta, \partial_3) V(x_1\chi_1, x_2\chi_2, x_3'\chi_{\ell\nu}^{-}\mathfrak{z})\, dx_3'$$

$$= \operatorname{sign}\left[\Gamma\left(\frac{1-\nu}{2} - c_-\right)\Gamma\left(\frac{1-\nu}{2} + c_-\right) V(x_1\chi_1, x_2\chi_2, x_3\chi_{\ell\nu}^{+}\zeta)\right) \tag{11.30a}$$

$$\frac{1}{(\ell+\nu)!}\int \mathcal{B}_{\ell\nu}^{\prime +}(x_{33'}; \zeta, \partial_3) V(x_1\chi_1, x_2\chi_2, x_3'\chi_{\ell\nu}^{\prime -}\mathfrak{z})\, dx_3' \tag{11.30b}$$

$$= \operatorname{sign}\left[\Gamma\left(\frac{1+\nu}{2} - c_-\right)\Gamma\left(\frac{1+\nu}{2} + c_-\right)\right] V(x_1\chi_1, x_2\chi_2, x_3\chi_{\ell\nu}^{\prime +}\zeta).$$

The sign factors in (11.30) appear in accord with the definition of N_ℓ in the cut c-plane (see Eq. (10.21) and subsequent remark). To establish (11.28) we apply the integral operator with kernel

$$\frac{(-1)^\nu \ell!^2}{(\ell+\nu)!^2} (\nabla_3 \partial_\zeta)^\nu \mathcal{B}'^{+}_{\ell\nu}(x_{33'}; \zeta, \partial_3)$$

to both sides of (11.24) and integrate the left hand side by parts. Using (11.29) we obtain

$$\frac{(2h+\ell-2)_\nu}{(h+\ell-1)_\nu} \int \mathcal{B}_{\ell\nu}(x_{33'}; \zeta, \partial_3) V(x_1\chi_1, x_2\chi_2, x_3'\chi_{\ell\nu}^- \mathfrak{z})\, dx_3' = \tag{11.31}$$

$$= \frac{(-1)^\nu \ell!^2}{(\ell+\nu)!^2} \operatorname{sign}\left[\left(\tfrac{1-\nu}{2}+c_-\right)_\nu\right] (\nabla_3 \partial_\zeta)^\nu \int \mathcal{B}'^{+}_{\ell\nu}(x_{33'}; \zeta, \partial_3) V(x_1\chi_1, x_2\chi_2, x_3'\chi'^{-}_{\ell\nu} \mathfrak{z})\, dx_3'$$

In order to obtain (11.28) it remains to substitute (11.30) in (11.31) taking into account

$$\operatorname{sign}\left[\frac{\Gamma\left(\frac{1-\nu}{2}-c_-\right)\Gamma\left(\frac{1+\nu}{2}+c_-\right)}{\Gamma\left(\frac{1+\nu}{2}-c_-\right)\Gamma\left(\frac{1-\nu}{2}+c_-\right)}\right] = (-1)^\nu$$

We note that although $\mathcal{B}'^{+}_{\ell\nu}$ is only defined as a bilinear form on the invariant subspace $F'_{\ell\nu}$ of $C'^{-}_{\ell\nu}$, Eq. (11.31) makes sense as an operator equation on $C'^{-}_{\ell\nu}$.

We can use (11.24) and (11.28) in order to derive similar identities for the "conformal Fourier transform" $F_\chi(x,\mathfrak{z})$, defined by (11.1). For example, applying the differential operator $(\mathfrak{z}\cdot\nabla)^\nu$ to both sides of the equation

$$F_{\chi^-_{\ell\nu}}(x,\mathfrak{z}) = \frac{1}{2}\iint f(x_1,x_2)\, V(x_1\tilde{\chi}_1 \cdot x_2\tilde{\chi}_2 ; x\,\chi^-_{\ell\nu}\,\mathfrak{z})\, dx_1\, dx_2$$

(which is (11.1) analytically continued in c (for $c_- \neq 0$)), and using (11.24)(with χ_i replaced by $\tilde{\chi}_i$, $i = 1, 2$) we obtain

$$(\mathfrak{z}\cdot\nabla)^\nu F_{\chi^-_{\ell\nu}}(x,\mathfrak{z}) = \operatorname{sign}\left[\left(\tfrac{1-\nu}{2} - c_-\right)_\nu\right] F_{\chi'^-_{\ell\nu}}(x,\mathfrak{z}) \,. \tag{11.24'}$$

Similarly, as a consequence of (11.28), we find the relation

$$\frac{\ell!}{(\ell+\nu)!}(\nabla\cdot\partial_\zeta)^\nu F_{\chi'^+_{\ell\nu}}(x,\zeta) = \operatorname{sign}\left[\left(\tfrac{1-\nu}{2} - c_-\right)_\nu\right] \frac{(2h+\ell-2)_\nu}{(h+\ell-1)_\nu} F_{\chi^+_{\ell\nu}}(x,\zeta) \tag{11.28'}$$

11.C. Tensor product representations and Clebsch-Gordan expansions for distributions

As already noted (see conclusion 10.8) the expansion formulae (10.41), (11.1) and (11.12) are valid for all square integrable functions. That includes the 1-particle irreducible Green functions to be considered in Chapter V below. We shall, however, also need to apply the intertwining differential operators of Secs. 6 and 11.8 to such functions. That is indeed possible if we regard them as distributions and use the notion of derivative of a distribution. The above expansion formulae (and related equations) can be extended from the space of test functions S_{κ_0} to the dual space of distributions in the line sketched below.

Consider the space S'_{κ_0} of continuous linear functionals on the test function space S_{κ_0}, defined in Sec. 11A. We shall use the distribution theoretic notation for the functionals Φ of S'_{κ_0} :

$$(\Phi, f) = \iint \Phi(x_1, x_2) f(x_1, x_2)\, dx_1 dx_2 \qquad (f \in S_{\kappa_0}) . \tag{11.32}$$

We define the dual T' to the representation $T = T_{\chi_1} \otimes T_{\chi_2}$ (acting in S_{κ_0}) by

$$(T'(g)\Phi, f) = (\Phi, T(g)f) \tag{11.33}$$

(the right-hand side is a continuous linear functional on S_{κ_0} $(\ni f)$ which is denoted by $T'(g)\Phi$). The representation T' of $G_{(ex)}$ can be regarded as an extension of the representation $\tilde{T} = T_{\tilde{\chi}_1} \otimes T_{\tilde{\chi}_2}$. Indeed, it is easily verified that $S_{\tilde{\kappa}_0} \subset S'_{\kappa_0}$ and that the action of T' , defined by (11.33), coincides with $\tilde{T}$ on $S_{\tilde{\kappa}_0}$ (where $S_{\tilde{\kappa}_0}$ is S_{κ_0} with $\chi_1 \to \tilde{\chi}_1, \chi_2 \to \tilde{\chi}_2$) - cf. Proposition 3.6.

Consider the topological vector space C_Σ of (test) functions $F_\chi(x)$ which can be presented in the form (11.1) with $f(x_1, x_2) \in S_{\kappa_0}$. We shall write

$$C_\Sigma = \oint^{\oplus} d\chi\, C_\chi \tag{11.34}$$

indicating that $F_\chi(x) \in C_\chi$ (for fixed χ). Regarded as a function of $\chi = [\ell, c]$ the right hand side of (11.1) allows an analytic continuation in

c (off the imaginary axis) because of the vanishing of $f(x_1,x_2)$ and its derivatives for coinciding arguments. It goes fast to zero for $|\Im m\, c| \to \infty$, because of the infinite differentiability of the test functions f.
A precise intrinsic charactic of the space C_Σ is desirable (in terms of the properties of $F_\chi(x)$ regarded as functions of two (sets of) variables x and χ), but we are not prepared to give such a characteristic here.

Let C'_Σ be the dual space of (continuous) linear functionals on C_Σ, which will be written symbolically as

$$\Phi(F) = \oint\!\!\!\!\!\sum d\chi \int dx\, \Phi_{\tilde{\chi}}(x)\, F_\chi(x) \qquad \left(F_\chi(x) \in C_\Sigma\ ,\ \Phi_{\tilde{\chi}}(x) \in C'_\Sigma\right) \tag{11.35}$$

Inserting for $f(x_1,x_2)$ in (11.32) its partial wave expansion (10.12) we can define a mapping from S'_{K_0} into C'_Σ ; formally

$$\Phi_{\tilde{\chi}}(x) = \int dx_1 \int dx_2\, \phi(x_1,x_2)\, V(x_1\chi_1, x_2\chi_2, x\tilde{\chi}) \qquad (\Phi_{\tilde{\chi}} \in C'_\Sigma) \tag{11.36}$$

The corresponding formula (11.1) for the test functions ensures that the above mapping is actually onto and has an inverse which can be written symbolically in the form

$$\Phi(x_1,x_2) = \oint\!\!\!\!\!\sum d\chi \int dx\, \Phi_{\tilde{\chi}}(x)\, V(x_1\tilde{\chi}_1, x_2\tilde{\chi}_2, x\chi) \tag{11.37}$$

Thus one indeed has a distribution theoretic counterpart of the tensor product expansion formulas (11.1) and (11.12).

V. DYNAMICAL DERIVATION OF VACUUM OPERATOR PRODUCT EXPANSION IN EUCLIDEAN CONFORMAL QUANTUM FIELD THEORY

12. Renormalizable models of self-interacting scalar fields. Dynamical equations for Green functions

12.A. A 6-dimensional model. Euclidean Green functions. Generating functionals.

The simplest model of a renormalizable self-interacting field $\varphi(x)$ is given by the interaction Lagrangian $L_I(x) = - g/3! \; :\varphi^3(x):$ in six space time dimensions. Although this model is manifestly sick (since the corresponding classical Hamiltonian is not positive definite) it can (and does) serve as a testing ground for various quantum field theoretic techniques (apart from its role in the work [S6,M2] which we are going to review, it presents the simplest example of a theory with asymptotic freedom -- see, e.g. [M1]). We shall indicate at the end of this section how one might modify the model, in order to cure it from its obvious disease.

Having in mind models in different number of dimensions, we shall work in a general framework of 2h-dimensional space time (2h=2, 3, 4, ...).

In what follows we shall mostly deal with Euclidean Green functions also called Schwinger functions (cf. [S7, O3]):

$$s(\underset{\sim}{x}_1\sigma_1, \dots \underset{\sim}{x}_n\sigma_n) = \tau(i\sigma_1 \underset{\sim}{x}_1, \dots, i\sigma_n \underset{\sim}{x}_n) \tag{12.1}$$

where $\tau(x_1,\dots,x_n) = \langle T\varphi(x_1)\dots\varphi(x_n)\rangle_0$ (φ is the interacting Heisenberg field).

One can define connected, one-particle irreducible (IPI) etc. Green functions without recourse to perturbation theory. The most compact way to do that is in terms of generating functionals (see e.g. [S6, C2]).

Let $J(x)$ be a scalar external source and let $\mathcal{S}(J)$ be the generating functional for the s-functions

$$\mathcal{S}(J) = 1 + \sum_{n=1}^{\infty} \frac{1}{n!} \int\dots\int dx_1 \dots dx_n \; s(x_1\dots x_n)\, J(x_1)\dots J(x_n)$$
$$\equiv \langle \exp \int J(x)\,\phi(x)\,dx \rangle_0 \quad ; \quad dx = d^{2h}x \tag{12.2}$$

($\phi(x)$ is by definition the Euclidean field). The generating functional $\mathcal{G}(J)$ of the connected (Euclidean) Green functions $G(x_1 \ldots x_n)$ is defined by

$$\mathcal{f}(J) = e^{\mathcal{G}(J)} \quad , \quad s(x) - G(x) = 0 \tag{12.3}$$

The source $J(x)$ is associated with a classical (Euclidean) field $\phi_c(x)$ by

$$\phi_c(x) = \frac{\delta \mathcal{G}}{\delta J(x)} = \mathcal{f}^{-1}(J) \frac{\delta \mathcal{f}}{\delta J(x)} \quad . \tag{12.4}$$

The generating functional for the 1PI Green functions (or <u>proper vertex functions</u>) $\Gamma(x_1 \ldots x_n)$ is given by the Legendre transformation [C2]

$$\begin{aligned} \Gamma(\phi_c) &= \mathcal{G}(J) - \int dx\, J(x)\phi_c(x) \\ &= \sum_{n=2}^{\infty} \frac{1}{n!} \int \ldots \int dx_1 \ldots dx_n\, \Gamma(x_1 \ldots x_n)\, \phi_c(x_1) \ldots \phi_c(x_n) \quad . \end{aligned} \tag{12.5}$$

To obtain the right-hand side of (12.5) we express $J(x)$ in terms of $\phi_c(x)$ from (12.4).

12.B. Graphical notation. 1i - and 2i - kernels

In order to write down the (renormalized) equations for the model under consideration we shall need some additional auxiliary notions (cf. [F7, S6, M2, F8]).

We introduce the <u>amputated</u> (connected) Green functions

$$A(x_1 \ldots x_n) = \int \ldots \int dy_1 \ldots dy_n\, G^{-1}(x_1 - y_1) \ldots G^{-1}(x_n - y_n)\, G(y_1 \ldots y_n) \,, \tag{12.6}$$

where

$$G^{-1} * G\,(x_1 - x_2) \equiv \int dx\, G^{-1}(x_1 - x)\, G(x - x_2) = \delta(x_1 - x_2) \tag{12.7}$$

(We use alternately the notation $G(x_1, x_2)$ and $G(x_1 - x_2)$ for the 2-point function; that is legitimate, because of translation invariance). We have

$$A(x_1, x_2) = -\Gamma(x_1, x_2) = G^{-1}(x_1 - x_2) \;,\; A(x_1 x_2 x_3) = \Gamma(x_1 x_2 x_3) \quad . \tag{12.8}$$

We shall use the graphical notation

$G(x_1 \ldots x_n) =$ 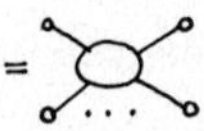 ; $G(x_1, x_2) =$ 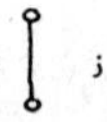 ; $G^{-1}(x_1, x_2) =$ 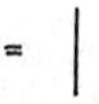 (12.9a)

$A(x_1 \ldots x_n) =$  ; $\Gamma(x_1 \ldots x_n) =$  (12.9b)

The 1PI amplitude for the channel $12 \to 3\ldots n$ $(n \geqslant 4)$ is defined by

$$A_{1i}(x_1 x_2 ; x_3 \ldots x_n) = A(x_1 \ldots x_n) - \int dy_1 \int dy_2\, A(x_1 x_2 y_1)\, G(y_1 y_2)\, A(y_2 x_3 \ldots x_n). \qquad (12.10a)$$

or graphically

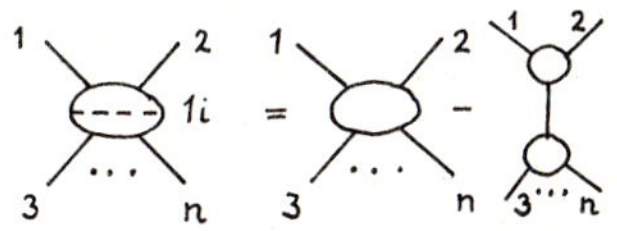

(12.10b)

We define the Bethe-Salpeter (BS) kernel

$B\,(x_1\; x_2\; ;\; x_3\; x_4) =$ 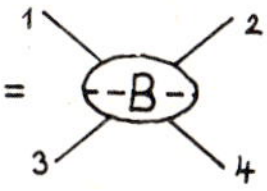

as the solution of the (integral) BS equation

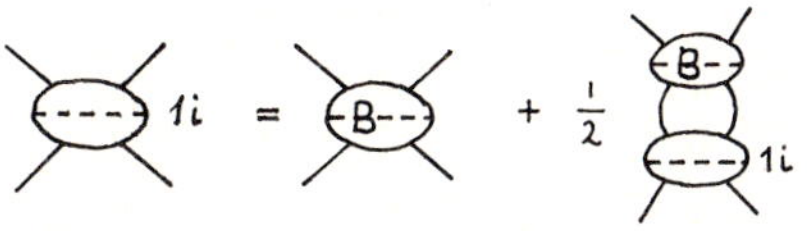

(12.11)

(The factor 1/2 in the right-hand side is necessary because of the symmetry of the theory of a single neutral scalar field). Finally we introduce the "2-particle irreducible" kernel for the channel $12 \to 3\ldots n$ $(n \geqslant 5)$ by induction in n:

$A_{2i}(x_1 x_2 ; x_3 x_4 x_5) \equiv$ 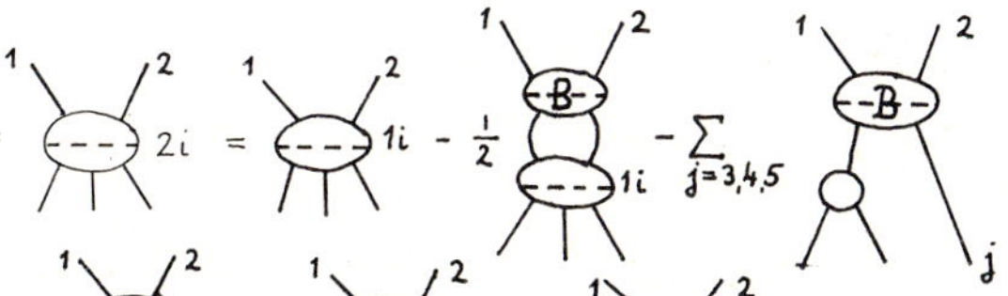

(12.12a)

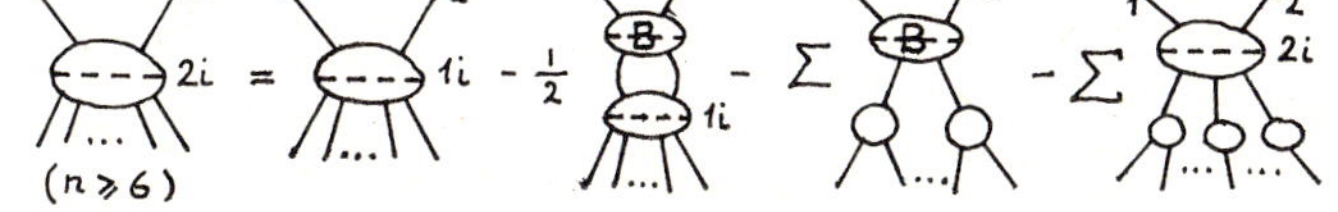

(12.12b)

The first sum in the right hand side of Eq. (12.12b) (and in (12.12a)) is spread over all $2^{n-3} - 1$ partitions of the set of external lines $3\ldots n$ into two non-empty subsets. The second sum involves all splittings of these lines into k non-empty subsets $(k = 3, \ldots, n-3)$.

12.C. Dynamical equations

Stress energy tensor. Ward identities

The dynamical equations can be written either in terms of the connected Green functions G(cf. [S6, M2]) or in terms of the proper vertex functions (cf. [F7, F8]) the two forms being equivalent. We shall adopt here the latter form, writing however, the equation for the 2-point function in a way suggested in [M6, M2].

In the Gell-Mann-Low limit (in which the renormalization constant $Z_1 = 0$ *)) the dynamical equations for the vertex functions have the form

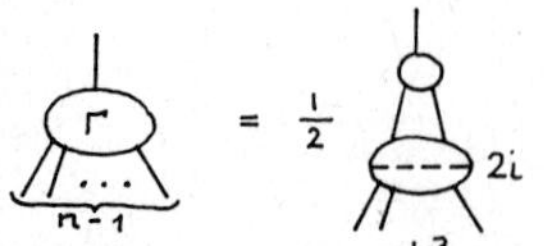

$n \geq 4$ (12.13)

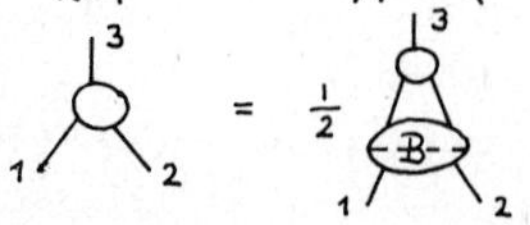

(12.14)

$$(x_1-x_2)_\nu \frac{\partial}{\partial x_{1\mu}} G^{-1}(x_1-x_2) =$$ 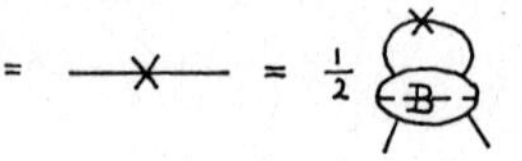 $, \quad (\nu \neq \mu)$ (12.15a)

The last equation can be written equivalently in the form

$$(x_1-x_2)_\nu \frac{\partial}{\partial x_{1\mu}} G(x_1-x_2) =$$ 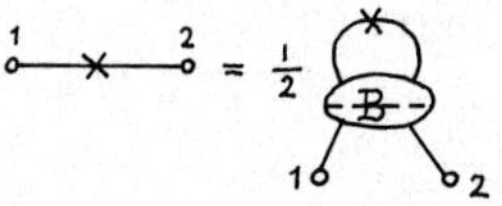 $(\nu \neq \mu)$ (12.15b)

and external lines have been attached to the BS kernel in the right hand side (which amounts to the inverse operation of (12.6)).

The bootstrap form of Eqs. (12.14) (12.15) is peculiar to the Gell-Mann-Low limit theory. In general (away from that limit), there is an inhomogeneous term in the right-hand side of (12.14) and both equations require subtractions in momentum space or multiplication by $(x_1-x_2)_\mu$ in coordinate space. Masses and coupling constants appear in such an approach as initial conditions (cf. ref. [M2]).

It is convenient to use an alternative form of Eq. (12.15) involving the stress energy tensor $\Theta_{\mu\nu}(x)$ (see ref. [M6]). Let $G_{\mu\nu}(x_1 x_2 ; x_3)$ be the Euclidean region continuation of $\langle T\varphi(x_1)\varphi(x_2)\Theta_{\mu\nu}(x_3)\rangle_0$ and let

$$\Gamma_{\mu\nu}(x_1 x_2 ; x_3) = \int dy_1 \int dy_2 \, G^{-1}(x_1-y_1)\, G^{-1}(x_2-y_2)\, G_{\mu\nu}(y_1 y_2 ; x_3)$$

---

*) We are using the convention of ref. [S6], according to which the renormalized field operator $\varphi(x)$ and coupling constant g are related to the corresponding unrenormalized quantities φ_u and g_u by

$$\varphi(x) = Z_1^{-\frac{1}{2}} :\varphi_u(x): \; ; \quad :\varphi_u(x): = -i \frac{\delta}{\delta J(x)} \langle T\exp(i\int \varphi_u(x) J(x) dx)\rangle_0^{-1} T\exp(i\int\varphi_u(x)J(x)dx)\Big\}_{J=0}$$

It should be noted that Eqs. (12.13) - (12.15) do not contain any parameter (like coupling constant), but just relate Green functions among themselves.

be the corresponding vertex function. It is assumed the $G_{\mu\nu}$ and $\Gamma_{\mu\nu}$ satisfy the following (equivalent between each other) Ward-Takahashi identities:

$$\nabla_3^\lambda G_{\lambda\mu}(x_1 x_2 ; x_3) = \sum_{i=1}^{2} [\delta(x_3-x_i)\nabla_{i\mu} G(x_1-x_2) - a\, G(x_1-x_2)\nabla_{i\mu}\delta(x_3-x_i)] \quad (12.17a)$$

$$\nabla_3^\lambda \Gamma_{\lambda\mu}(x_1 x_2 ; x_3) = -\sum_{i=1}^{2} [\delta(x_3-x_i)\nabla_{i\mu} G^{-1}(x_1-x_2) + (1+a) G^{-1}(x_1-x_2)\nabla_{i\mu}\delta(x_3-x_i)] \quad (12.17b)$$

The last (Schwinger) term can be eliminated by multiplying both sides of each of the equations (12.17) by $(x_1 - x_3)_\nu$ $(\nu \neq \mu)$ and integrating over x_3 with the result

$$\int dx\, \nabla^\lambda G_{\lambda\mu}(x_1 x_2 ; x)(x_1-x)_\nu = (x_1-x_2)_\nu \frac{\partial}{\partial x_{2\mu}} G(x_1-x_2) \quad (12.18a)$$

$$\int dx\, \nabla^\lambda \Gamma_{\lambda\mu}(x_1 x_2 ; x)(x_1-x)_\nu = (x_1-x_2)_\nu \frac{\partial}{\partial x_{1\mu}} G^{-1}(x_1-x_2) \quad (12.18b)$$

This form of the identity has the advantage to be independent of the arbitrary constant a .

The set of equations (12.13) - (12.15) is equivalent to the set in which (12.15) is replaced by

$\Gamma_{\mu\nu}(x_1 x_2 ; x_3) \equiv$ 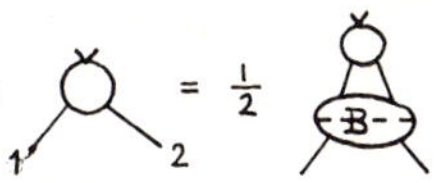 (12.19a)

$G_{\mu\nu}(x_1 x_2 ; x_3) \equiv$ 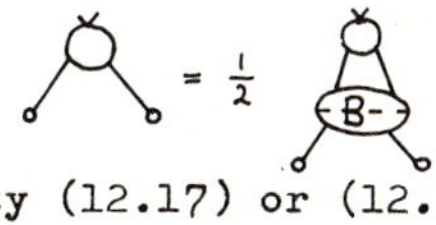 (12.19b)

and the Ward identity (12.17) or (12.18) is assumed to hold. As a consequence we obtain an infinite set of additional integral equations for the n + 1 point functions $\Gamma_{\mu\nu}(x_1 \ldots x_n ; x)$ involving the stress energy tensor:

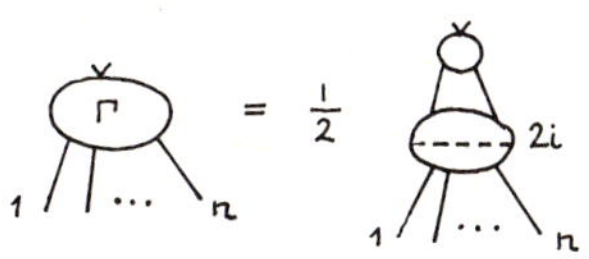 $n = 3, 4, \ldots$ (12.20)

They also satisfy Ward identities of the type (12.17). To be consistent with the scale invariance of the Gell-Mann-Low limit theory, we have to require that $\Theta_{\mu\nu}$ is traceless, so that

$$\Gamma_{\mu\mu}(x_1 \dots x_n ; x) = 0 = G_{\mu\mu}(x_1 \dots x_n ; x) \qquad n = 2,3,\dots \tag{12.21}$$

12.D. A more realistic model

Although the above dynamical equations are derived from a renormalizable Lagrangian in 6 space time dimensions, their final form (12.13) - (12.20) makes sense for an arbitrary number 2h of dimensions.

Here we shall indicate how one can fit the more realistic model of a charged (pseudo) scalar field with a quartic interaction in four space-time dimensions

$$\mathcal{L}(\varphi(x), \varphi^*(x)) = :\nabla_\mu \varphi^* \nabla^\mu \varphi: - \frac{\lambda}{2} :(\varphi^* \varphi)^2: \tag{12.22}$$

into the above framework.

The clue lies in the observation (made, e.g., by Symanzik) that the model given by (12.22) can equivalently be described by the Lagrangian

$$\mathcal{L}(\varphi(x), \varphi^*(x); B(x)) = :\nabla_\mu \varphi^* \nabla^\mu \varphi: + \frac{1}{2} :B^2: - \sqrt{\lambda} :\varphi^* \varphi B: \tag{12.23}$$

of a system of two fields, φ and B , with a cubic interaction. Indeed, varying $\mathcal{L}(\varphi, \varphi^*; B)$ with respect to B , we find the algebraic "equation of motion" $B = \sqrt{\lambda} :\varphi^* \varphi:$ which reduces (12.23) to (12.22). In a canonical perturbation theory the propagator corresponding to the field B would be a δ-function. On the other hand, the topological structure of Feynman diagrams in this model is the same as in a theory with Yukawa coupling of a charged field φ and a neutral scalar field B (which will be represented graphically by a cut line ---).

Without going into the details of the Green function formulation of this model, we notice that it will involve (a priori)

four types of vertex functions

$$= g_1 V(x_1, -c_\varphi ; x_2, -c_\varphi ; x_3, -c_B) \tag{12.24}$$

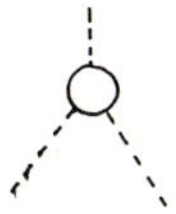

$$= g_0 V(x_1, -c_B ; x_2, -c_B ; x_3, -c_B) \tag{12.25}$$

$$= G^{(1)}_{\mu\nu}(x_1, x_2 ; x_3) \tag{12.26}$$

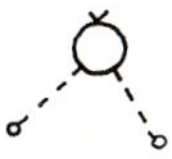

$$= G^{(2)}_{\mu\nu}(x_1, x_2 ; x_3) \tag{12.27}$$

The bootstrap equation for the charged propagator can be equivalently obtained [M6] from the corresponding equation for the current-field 3-point function

$$\langle T\, \varphi(x_1)\, \varphi^*(x_2)\, j_\mu(x_3) \rangle_0 \Leftrightarrow G_\mu(x_1 x_2 ; x) = \tag{12.28}$$

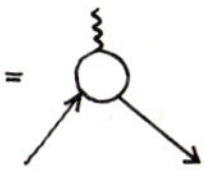

which satisfies the Ward identity

$$\nabla_3^\mu G_\mu(x_1 x_2 ; x_3) = e\, G(x_1 - x_2)[\delta(x_1 - x_3) - \delta(x_2 - x_3)] \tag{12.29}$$

(e being the electric charge carried by $\varphi^*(x)$).

13. Invariance and invariant solutions of the dynamical equations. Conformal partial wave expansion for the Euclidean Green functions

13.A. Euclidean conformal invariance of the equations .

As already noted, all equations of Sec. 12 only relate Green functions among themselves. They involve no parameters (in particular, no dimensional parameters) and are manifestly scale invariant. Indeed, if we ascribe to the field φ in 2h dimensions a scale dimension $d_\varphi = h + c$ (c real), then the Green functions G and Γ would have the following transformation properties under dilation:

$$G(x_1 \ldots x_n) \to \rho^{n(h+c)} G(\rho x_1, \ldots, \rho x_n) \tag{13.1a}$$

$$\Gamma(x_1, \ldots, x_n) \to \rho^{n(h-c)} \Gamma(\rho x_1, \ldots, \rho x_n) \quad , \qquad \rho > 0 \; ; \tag{13.1b}$$

in particular,

$$G^{-1}(x_1 - x_2) = -\Gamma(x_1, x_2) \to \rho^{2(h-c)} G^{-1}(\rho x_1 - \rho x_2) \tag{13.1c}$$

(the canonical dimension for a spinless field φ is obtained for $c = -1$). Eqs. (12.13) - (12.15) are obviously invariant under the substitution (13.1). The Ward identities (12.17) and (12.30) imply that the scale dimension of the stress energy tensor $\Theta_{\mu\nu}$ and of the conserved current j_μ are

$$d_\Theta = 2h \quad , \quad d_j = 2h-1 \tag{13.2}$$

respectively. The dimension d_B of the field B in the model, considered in Sec. 12.D, can be ascribed indepemdently.

It turns out that the dynamical equations are invariant under the Euclidean conformal group $O^{\uparrow}(2h+1,1)$ which, as we know from the Bruhat decomposition (Sec. 1.B), can be generated by (Euclidean) Poincaré transformations, dilatations and the conformal inversion (cf. (1.29)):

$$Rx = -\frac{x}{x^2} \qquad (x^2 = x_1^2 + \ldots + x_{2h}^2) \tag{13.3}$$

The transformation law of Green functions under the inversion (13.3) is summarized by the following rules for Euclidean fields:

$$U(R)\phi(x)U(R)^{-1} = (x^2)^{-h-c}\phi(Rx), \tag{13.4a}$$

$$U(R)J_\mu(x)U(R)^{-1} = (x^2)^{-2h+1} r_{\mu\nu}(x) J_\nu(Rx) \quad , \tag{13.4b}$$

$$U(R)\Theta_{\mu\nu}(x)U(R)^{-1} = (x^2)^{-2h} r_{\mu\mu'}(x) r_{\nu\nu'}(x)\, \Theta_{\mu'\nu'}(x) \; , \tag{13.4c}$$

where $r_{\mu\nu}(x)$ is given by (2.30b):

$$r_{\mu\nu}(x) = x^2 \nabla_\mu (Rx)_\nu = \frac{2x_\mu x_\nu}{x^2} - \delta_{\mu\nu} . \tag{13.5}$$

The so-called special conformal transformations are given by a translation, sandwiched between two inversions. The R-invariance of the dynamical equations follows from the covariance law for the volume element

$$dRx = (x^2)^{-2h} \det (r_{\mu\nu})\, dx = (x^2)^{-2h} dx . \tag{13.6}$$

13.B. Conformal invariant 2- and 3- point functions

We shall study in the rest of the paper conformal invariant solutions of the dynamical equations for the models described in Sec. 12. If the Gell-Mann-Low limit is ultraviolet stable (as usually assumed in this approach) then the conformal invariant solution would provide the small distance behavior of Green functions in a more realistic theory with positive mass particles.

The invariance property of the solution allows one to determine the 2- and 3-point functions up to a constant factor without actually solving the equations. Before writing down the corresponding expressions, we shall make a remark about the freedom in the choice of normalization.

In a canonical field theory the field operators are normalized in such a way that the residue in the pole of the 2-point function is one. In a scale invariant theory with anomalous dimensions the 2-point function has no pole and there is no unique choice of field normalization. A multiplicative change $\varphi(x) \to \kappa\varphi(x)$ in the field (where $\kappa = \kappa(c)$ is some function of the dimension) leads to the following transformation law for Green functions

$$\begin{aligned} &G(x_1 \dots x_n) \to \kappa^n G(x_1 \dots x_n) \\ &\Gamma(x_1 \dots x_n) \to \kappa^{-n} \Gamma(x_1 \dots x_n) \\ &G_{\mu\nu}(x_1 \dots x_n ; x) \to \kappa^n G_{\mu\nu}(x_1 \dots x_n ; x) \end{aligned} \tag{13.7}$$

[The normalization of $\Theta_{\mu\nu}$ is fixed by the Ward identity (12.17) .]

Thus we can choose the normalization of the 2-point function of φ (and B) according to convenience; only the relative normalization of the 2- and 3-point function will have a physical significance.

We shall choose the normalized invariant 2-point function for a fundamental scalar field φ to be (4.19) with normalization (5.25):

$$G_{[0,c]}(x) \equiv G(x_1 - x_2) = (2\pi)^{-h} \frac{\Gamma(c+h)}{\Gamma(-c)} \left(\frac{2}{x_{12}^2}\right)^{h+c} ; \quad x_{12} = x_1 - x_2 . \tag{13.8}$$

With this normalization the Fourier transform of G is (cf.(5.23a)):

$$G(p) = \int e^{-ipx} G(x)\, dx = \left(\tfrac{1}{2}p^2\right)^c \tag{13.9}$$

The inverse Green function $G^{-1}(x_{12})$ is obtained from (13.8) by changing the sign of c. We shall say that the field φ is fundamental, if its dimension parameter $c = c_\varphi$ satisfies the inequalities

$$-1 \leqslant c_\varphi < 0 \tag{13.10}$$

For a fundamental field φ the two-point function G corresponds to a positive definite Wightman function (cf. the corresponding formula for the composite fields (5.31))

$$\begin{aligned} w(p) &= -i\,\theta(p_0)\left[G(\underline{p}, -ip_0 + 0) - G(\underline{p}, -ip_0 - 0)\right] \\ &= -2\theta_+(p) \sin \pi c \left(\tfrac{1}{2}p_M^2\right)^c \end{aligned} \tag{13.11}$$

where $\theta_+(p) = \theta(p_0)\theta(p_M^2)$, $p_M^2 = p_0^2 - \underline{p}^2$. (The subscript M stands for real Minkowski space vectors and scalar products). If we multiply φ by $\kappa(c) = [2^c\Gamma(-c)]^{1/2}$ we will obtain an elementary field (cf. Sec. 5.D) with positive definite Wightman function for all $c \geqslant -1$ which coincides with the 2-point function of a free 0-mass field for c=-1.

We shall also need in what follows the conformal invariant 2-point function of an arbitrary rank ℓ symmetric traceless tensor field $O_{\mu_1 \ldots \mu_\ell}(x)$ of dimension h+c, c=c(ℓ) . Again we shall use the normalized 2-point function (4.19) with normalization (5.25) ($\chi = [\ell, c]$):

$$G_\chi(x_{12}; \zeta_1, \zeta_2) = \frac{n_0(\chi)}{(2\pi)^h} \left(\frac{2}{x_{12}^2}\right)^{h+c} \left(-\zeta_1\, r(x_{12})\, \zeta_2\right)^\ell \tag{13.12}$$

With this normalization the momentum space 2-point function becomes (see (5.23a)):

$$G_\chi(p) = \left(\tfrac{1}{2}p^2\right)^c \sum_{s=0}^{\ell} \alpha_s(c)\, \Pi^{\ell s}(p) \tag{13.13}$$

where (cf. Sec. 5.B)

$$\Pi^{\ell s}(p;\mathfrak{z}_1\cdot\mathfrak{z}_2) = \frac{(s+h-1)_{\ell-s}}{(2h+2s-2)_{\ell-s}}\binom{\ell}{s}\left(2\frac{p\cdot\mathfrak{z}_1\, p\cdot\mathfrak{z}_2}{p^2}\right)^{\ell-s}\Pi^{ss}(p;\mathfrak{z}_1,\mathfrak{z}_2)$$

$$\Pi^{ss}(p;\mathfrak{z}_1,\mathfrak{z}_2) = \frac{s!(h-1)_s}{(2h-3)_{2s}}\left(-2\frac{p\cdot\mathfrak{z}_1\, p\cdot\mathfrak{z}_2}{p^2}\right)^{s} C_s^{h-\frac{3}{2}}\left(1-\frac{p^2\mathfrak{z}_1\cdot\mathfrak{z}_2}{p\cdot\mathfrak{z}_1\, p\cdot\mathfrak{z}_2}\right) \tag{13.14}$$

are the $SO(2h-1)_p$ projection operators,

$$\alpha_s(c) = \frac{(h+c-1)_s}{(h-c-1)_s}$$

We know (see (4.11)) that the inverse 2-point function is given by

$$G_\chi^{-1}(x_{12}) = G_{\tilde{\chi}}(x_{12}) \tag{13.15}$$

In order to be able to handle the most general situation (including the model described in Sec. 12D) we shall write down the 3-point function for three different spinless fields with dimension parameters c_1, c_2 and c_3. They are

$$G(x_1,x_2,x_3) = gV(x_1c_1;x_2c_2;x_3c_3) \tag{13.16}$$

$$\Gamma(x_1,x_2,x_3) = gV(x_1,-c_1;x_2,-c_2;x_3,-c_3), \tag{13.17}$$

where (see (11.4a)):

$$V(x_1c_1;x_2c_2;x_3c_3) = V(x_1\chi_1;x_2\chi_2;x_3\chi_3) \quad , \chi_i = [0,c_i] .$$

The function V can be associated with an "infraparticle" triangular diagram (see [M9], [P1], [M7]) with scale invariant propagators $(\frac{1}{2}x_{ik}^2)^{-d_{ik}}$. The parameters d_{ik} satisfy the conservation of dimension law in each vertex of the diagram:

$$d_{ik} + d_{jk} = d_k = h+c_k \quad ; \quad (ijk) - \text{Permutation } (123) \tag{13.18}$$

We shall also need the 3-point functions for two scalar (or pseudoscalar) fields of dimension parameters c_1 and c_2, and a rank ℓ tensor field $O(x,\mathfrak{z})$ of dimension h+c. They are given by

$$G(x_1,x_2;x_3\mathfrak{z}) = g_\ell V(x_1c_1;x_2c_2;x_3,\chi,\mathfrak{z}), \tag{13.19}$$

$$\Gamma(x_1,x_2;x_3\mathfrak{z}) = g_\ell V(x_1,-c_1;x_2,-c_2;x_3,\chi,\mathfrak{z}), \tag{13.20}$$

where (see (10.4a)):

$$V(x_1c_1;x_2c_2;x_3\cdot\chi\cdot\mathfrak{z}) = V(x_1\chi_1;x_2\chi_2;x_3\cdot\chi\cdot\mathfrak{z}), \quad \chi_i = [0,c_i], \; i=1,2. \tag{13.21}$$

(We do not consider amputation of the external line associated with the tensor field O). The normalization factor $N_\ell = N_\ell(c_+c_-;c)$ is given by (11.21).

The normalization of the 3-point function of the stress energy tensor is fixed by the Ward identity (12.17).

We have $G(x_1, x_2; x_3\zeta) = 0$ for $c_1 \neq c_2$, and

$$G(x_1, x_2; x_3\zeta) = \frac{\Gamma(h+1)}{2(2\pi)^{2h}} \frac{\Gamma(h+c+1)}{(2h+1)\Gamma(-c)} \left(\frac{2}{x_{12}^2}\right)^{c+1} \left(\frac{4}{x_{13}^2 x_{23}^2}\right)^{h-1} (\zeta\cdot\lambda)^2 = \tag{13.22a}$$

$$= \frac{\Gamma(h-1)}{2(2\pi)^{2h}} \frac{\Gamma(h+c+1)}{(2h-1)\Gamma(-c)} \left(\frac{2}{x_{12}^2}\right)^{c+1} \left[(\zeta\cdot\nabla_1)^2 - \frac{2h}{h-1}(\zeta\cdot\nabla_1)(\zeta\cdot\nabla_2) + (\zeta\cdot\nabla_2)^2\right] \left(\frac{4}{x_{13}^2 x_{23}^2}\right)^{h-1} \tag{13.22b}$$

With this G the coefficient to the Schwinger term in (12.17) is $a = -\frac{h+c}{2h}$.
In verifying (12.17) we have used the relations

$$(2h-1)^{-1}(D\nabla_3)\left[(\zeta\cdot\nabla_1)^2 - \frac{2h}{h-1}(\zeta\cdot\nabla_1)(\zeta\cdot\nabla_2) + (\zeta\nabla_2)^2\right] f(x_{12}, x_{23}) = -\left[\nabla_1^2\left(\zeta\cdot\nabla_1 - \frac{h+1}{h-1}\zeta\cdot\nabla_2\right) + \nabla_2^2\left(\zeta\cdot\nabla_2 - \frac{h+1}{h-1}\zeta\cdot\nabla_1\right)\right] f(x_{13}, x_{23})$$

$$-\frac{1}{2}(2\pi)^{-h}\Gamma(h-1)\,\nabla_1^2 (2x_{13}^{-2})^{h-1} = \delta(x_{13})$$

13.C. Skeleton diagram expansion

Having constructed the physical propagator and (3-point) vertex function one can expand the n-point functions $\Gamma(x_1 \ldots x_n)$ $(n \geq 4)$ in terms of skeleton diagrams*). It is important to know for the self consistency of conformal invariance that the skeleton diagrams, as well as the graphs encountered in the bootstrap equations (12.14) (12.19) have no ultraviolet (or momentum independent infrared) divergences. Indeed, such divergences would have destroyed even the scale invariance of the Green functions. It was demonstrated in ref. [M7], that for a certain range of the parameters c the boson-fermion Yukawa interaction in 4 dimensions is divergence free (for the Gell-Mann-Low limit theory under consideration). The analysis of [M7] is trivially extended to the 2h dimensional models considered in Sec. 12. For the simplest φ^3 model the most stringent restriction on the scaling parameter $c = c_\varphi$ comes from the requirement that the 3-point functions $G(x_1, x_2, x_3)$ and $\Gamma(x_1, x_2, x_3)$ are both given by ordinary convergent integrals in momentum space. That leads to

$$-\frac{1}{3}h < c < \frac{1}{3}h \tag{13.23}$$

The 3-point function $G_{\mu\nu}$ (see (13.22)) can be expressed in terms of a convergent p-space integral if $c < 0$, i.e. if the field is fundamental (see (13.10)). However, it can be continued analytically in c for $c > 0$

*) A skeleton diagram is a Feynman diagram in which internal lines correspond to dressed propagators and points are associated with physical vertices, and which contain no self-energy insertions or (3-point) vertex function corrections. Cf. [B4].

to cover the range (13.23), only the point $c=0$ being excluded.

Coming to the more realistic model envisaged in Sec. 12.D., we see that the existence of the 3-point functions (12.24) and (12.25) (complete and amputated on either leg) as convergent integrals in momentum space gives

$$2|c_\varphi| + |c_B| < h \tag{13.24}$$

and

$$-\tfrac{1}{3}h < c_B < \tfrac{1}{3}h \tag{13.25}$$

respectively. The convergence condition for the skeleton diagram

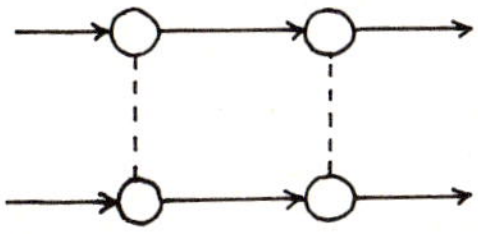

(13.26)

leads to

$$-\frac{h}{2} < c_\varphi < \frac{h}{2} \tag{13.27}$$

Assuming that φ is a fundamental field while B is a composite one (cf. Sec. 5.C) so that $c_\varphi < 0$, $c_B > 0$, we end up with the following complete set of inequalities

$$-\tfrac{1}{2}h < c_\varphi < 0 \quad , \quad 0 < c_B < \tfrac{1}{3}h \quad , \quad 0 < c_B - 2c_\varphi < h \tag{13.28}$$

which guarantee absence of divergences.

The skeleton expansion does not satisfy however the dynamical equations for all values of g and c. It turns out [M7, M6] that the entire infinite set of equations (12.13) - (12.19) will be satisfied provided that the two bootstrap equations (12.14) and (12.15) or (12.19), (12.18) are satisfied. Since both sides of (12.14) and of (12.19) are conformal invariant they have to be proportional to the 3-point functions (13.21) with $\ell = 0$ and 2, respectively. Thus, these bootstrap equations lead to coordinate independent transcendental equations for the two parameters g and c of the theory. As it could have been predicted these equations turn out to be equivalent to the Gell-Mann-Low equation for the coupling constant. That was verified by the ε-expansion method for the φ^3- theory in $6+\varepsilon$ dimensions in ref. [M1]. Unfortunately, this new version of the self-consistency equations does not seem any easier to handle. That is one reason why a new approach to the whole problem was attempted in [M2, D4] and is going to be pursued in what follows.

13.D. Conformal partial wave expansion

Eqs. (12.13) - (12.15) can be regarded as generalized (off-shell) unitarity equations. It is well known that in terms of the ordinary partial waves the (elastic) unitarity condition becomes an algebraic equation. It was demonstrated in ref. [M2] that the conformal extension of the partial wave analysis allows to solve the infinite set of dynamical equations for the φ^3- model.

Ordinary partial wave expansion is nothing else but the tensor product expansion of two irreducible (positive-energy) representations of the Poincaré group. Conformal partial wave expansion is by definition the tensor product expansion of two irreducible representations of the Euclidean conformal group (see Chapter IV).

The proper vertex function $\Gamma(x_1 x_2 \ldots x_n)$ ($n \geq 4$) considered as a function of the first two coordinates and integrated over the remaining n-2 coordinates with a "nice" test function f satisfies the square integrability condition

$$\int dx_1 \int dx_2 \int dy_1 \int dy_2 \, \bar{\Gamma}_f^{(n)}(x_1,x_2)\, G_{\chi_1}(x_1-y_1)\, G_{\chi_2}(x_2-y_2)\, \Gamma_f^{(n)}(y_1,y_2) < \infty \tag{13.29}$$

where

$$\chi_1 = [0, c_1] \;,\; \chi_2 = [0, c_2]$$

$$\Gamma_f^{(n)}(x_1,x_2) = \int \cdots \int dx_3 \ldots dx_n \, \Gamma(x_1 \ldots x_n)\, f(x_3 \ldots x_n) \tag{13.30}$$

in any order of the skeleton perturbation theory [D2]. (This is, however, not true for the 1-particle reducible diagrams, appearing in the right-hand side of Eq. (12.10b)). The integral in (13.29) is nothing else but the scalar product in the representation space of the tensor product of two irreducible complementary series representations χ_1 and χ_2 (as long as $-h < c_i < h$, $i=1,2$ which is certainly true if either (13.23) or (13.24)-(13.28) take place). If we assume in addition that (cf. (11.13))

$$|c_1| + |c_2| < h \qquad |\mathrm{Re}\, c_1| + |\mathrm{Re}\, c_2| < h \tag{13.31}$$

which is always true if the convergence conditions (13.23)-(13.28) are satisfied, then we can use the tensor product expansion formula (11.12) with respect to the first two arguments

$$\Gamma(x_1 \ldots x_n) = \sum\!\!\!\!\!\!\int d\chi \int dx\; V(x_1, -c_1;\, x_2, -c_2;\, x\tilde{\chi})\, \Gamma_\chi(x;\, x_3 \ldots x_n) \tag{13.32}$$

For $c_1 = c_2$ ($c_- = 0$) and Γ symmetric with respect to (x_1, x_2) the sum in (13.42) is over even values of ℓ only. The conformal partial wave Γ_χ is, conversely, expressed in terms of $\Gamma(x_1 \ldots x_n)$ by (cf.(11.1)):

$$\Gamma_\chi(x;\, x_3 \ldots x_n) = \frac{1}{2} \int dx_1 \int dx_2\; V(x_1, c_1;\, x_2 c_2;\, x\chi)\, \Gamma(x_1 \ldots x_n)\,. \tag{13.33}$$

In the special case of the 4-point function it follows from conformal invariance that the partial wave $\Gamma_\chi(x;\, x_3, x_4)$ is again proportional to V:

$$\Gamma_\chi(x;\, x_3, x_4) = \gamma(\chi)\, V(x_3, -c_3;\, x_4, -c_4;\, x\chi) \tag{13.34}$$

The conformal "Fourier transform" $\gamma(\chi)$ of the 4-point function $\Gamma(x_1 x_2 x_3 x_4)$ also depends on the dimension parameters c_i of the underlying fields, but not on the x's. The entire space-time dependence of the integrand in (13.32) is then given by a standard known function of the x's (the integral in x of the product of two V's).
Conversely, using (13.33) we can express the conformal partial wave $\gamma(\chi)$ in terms of Γ. We assume at this point that

$$c_1 - c_2 = c_3 - c_4 = 2c_- \tag{13.35}$$

According to Appendix F.2 the result is

$$\gamma(\chi) = \frac{\ell!\, b_\ell(-c_{12}, -c_{34}, c)}{2(2h-2)_\ell} \left(\frac{x_{34}^2}{2}\right)^{c_{34}} \int dx_1 \int dx_2 \left(\frac{x_{34}^2}{x_{12}^2}\right)^{h-\delta_{\tilde{\chi}}} \left(\frac{x_{12}^2}{2}\right)^{c_{12}} C_\ell^{h-1}\!\left(\frac{x_{12}\cdot x_{34}}{\sqrt{x_{12}^2 x_{34}^2}}\right) \Gamma(x_1 x_2 x_3 x_4)\,. \tag{13.36}$$

where $c_{ik} = \frac{1}{2}(c_i + c_k)$ and the factor b_ℓ is given by Eq. (F.7) of Appendix F.

A similar formula can be obtained by exchanging the roles of (x_1, x_2) and (x_3, x_4). The two expressions are consistent between each other because of the symmetry of $\Gamma(x_1 \ldots x_4)$ with respect to the substitution $(x_1 c_1;\, x_2 c_2) \rightleftarrows (x_3 c_3, x_4 c_4)$.

The symmetry property (11.8a) (written for the third argument) of the Clebsch-Gordan kernels implies the following relations between conformal partial waves

$$\Gamma_{\tilde{\chi}}(x;\, x_3 \ldots x_n) = \int dy\, G_{\tilde{\chi}}(x-y)\, \Gamma_\chi(y;\, x_3 \ldots x_n)\,, \tag{13.37}$$

$$\gamma(\tilde{\chi}) = \gamma(\chi) \tag{13.38}$$

13. E. Further expansion

Using the results of Sec. 10 one can continue the expansion process (13.32), expanding next the partial wave $\Gamma_\chi(x, x_3 \dots x_n)$ considered as a function of x, x_3, and so on. In order not to have too many indices we write temporarily $\Gamma(x\chi; \dots)$ in place of $\Gamma_\chi(x, \dots)$. In the language of Sec. 10, expansion formula (13.32) reads then ($\chi_0^* = \tilde{\chi}_0 = [0, -c]$, c real)

$$\Gamma(x_1 x_2; x_3 \dots x_n) = \sum_\ell \frac{1}{2\pi i} \int_{-i\infty}^{+i\infty} \rho(\ell, c) dc \int dx\, V(x_1 \tilde{\chi}_0 x_2 \tilde{\chi}_0, x\chi)^* \Gamma(x\chi | x_3 \dots x_n) \tag{13.39a}$$

The only nonvanishing contributions to the sum come from completely symmetric tensor representations ℓ of M. The partial waves

$$\Gamma(x\chi | x_3 \dots x_n) = \frac{1}{2} \int dx_1 dx_2 V(x_1 \chi_0 x_2 \chi_0; x\chi) \Gamma(x_1 x_2; x_3 \dots x_n) \tag{13.39b}$$

$V(\cdots x\chi)$ and $\Gamma(x\chi|\dots)$ are vectors in the $\mathcal{V}^\ell$. For such vectors $v_1^* v_2 = \langle v_1, v_2 \rangle$ is the scalar product in $\mathcal{V}^\ell$.

The formula is valid both for even and odd dimension 2h, provided the field dimension $d > \frac{1}{2}h$. [Otherwise there is an extra supplementary series contribution with $\chi = [0, 2c_0]$. It may be called 1-particle reducible and assumed subtracted in Γ; formula (13.39a) is then generally true].

Conformal invariance implies that also the partial waves are conformal invariant in the sense that, for $g \in G$

$$\{D^{\tilde{\chi}_0}(p_3) \dots D^{\tilde{\chi}_0}(p_n)\} D^\chi(p) \Gamma(x'\chi | x_3' \dots x_n') = \Gamma(x\chi | x_3 \dots x_n) \tag{13.40}$$

for $x_i' = g^{-1} x_i$, $x' = g^{-1} x$, $p_i = p(x_i, g)$, $p = p(x, g)$ defined by (1.27a).

The expression in {} is a scalar factor.

Let $\varphi(x_4 \dots x_n)$ be any test function and define $f(x x_3) = \int dx_4 \dots dx_n \varphi(x_4 \dots x_n) \Gamma(x\chi|x_3 \dots x_n)$. We make another regularity assumption which amounts to an a priori constraint on the strength of singularity of the partial wave at $x=x_3$. We assume that

$$(f,f) < \infty \qquad \text{(scalar product (10.43b))} \tag{13.41}$$

This generalizes hypothesis (13.29).

$f(x,x_3)$ transforms according to the Kronecker product of the class I complementary series representation χ_0 with a principal series representation $\chi = [\ell, c]$. We can again apply the conclusion and expansion formula (10.41) of Sec. 10, for even 2h. The Clebsch-Gordan kernels V are labelled by j,s. However since $\ell_0 = 0$, $\ell_0 \otimes \ell \simeq \ell$, and so $j = \ell$. We write $V^s(x\chi\, x_3\chi_0; x'\chi')$ in place of $V^{\ell s}(x\chi, x_3\chi_0; x'\chi')$. So the expansion for f is, reexpressed in terms of $\Gamma(x\chi|\dots)$

$$\Gamma(x\chi|x_3 \dots x_n) = \oint d\chi' \sum_s d(s)^{-1} \int dx' V^s(x\chi\, x_3\tilde{\chi}_0; x'\chi')^* \Gamma^s(x'\chi'\chi|x_4 \dots x_n) \tag{13.42}$$

with sum over all $s \in \hat{U}$ contained in $\ell' \in \hat{M}$ for $\chi' = [\ell', c']$.
The inversion formula is

$$\Gamma^s(x'\chi'\chi|x_4 \dots x_n) = \int dx\, dx_3 V^s(x\chi\, x_3\chi_0; x'\chi') \Gamma(x\chi|x_3 \dots x_n) \tag{13.43}$$

Eq. (13.42) may be inserted into expansion (13.39a) for the vertex-function. The procedure may then be repeated, i.e. we expand next $\Gamma^s(x'\chi'\chi|x_4 \dots)$ considered as a function of x' and x_4, and so on. At each step the number of x-space variables in the conformal partial wave (cpw.) is reduced by one.

We can also start expanding at the other end. The complex conjugate $\bar{\Gamma}(x_n \dots x_1)$ has the same conformal transformation law as $\Gamma(x_1 \ldots x_n)$ because $D^{\tilde{\chi}_0}(man) = |a|^{-h+c_0} = |a|^{-2h+d}$ is real. So we may apply expansion formula (13.39) to it. Taking the complex conjugate again we obtain a formula of the form

$$\Gamma(x_1 \ldots x_n) = \oint d\chi \int dx \, \Gamma(x_1 \ldots x_{n-2} | x\chi)^* \, V(x_{n-1} \chi_0^* x_n \chi_0^* ; x\chi)$$

and the conformal partial wave in this case is a "row vector" in $\mathcal{V}^\ell$ and has covariance property

$$\{D^{\chi_0}(p_1) \cdots D^{\chi_0}(p_{n-2})\} \, \Gamma(x_1' \ldots x_{n-2}' | x'\chi)^* D^\chi(p)^* = \Gamma(x_1 \ldots x_{n-2} | x\chi)^* \qquad (13.40')$$

(same notation as in (13.40))

The expansion process may be continued step by step, much as before.

The expansion processes starting at each end may be combined. At each step in the expansion process, the number of x-space variables in the cpw. is reduced by one. After sufficiently many steps it will then depend on only two such variables, say x and x'. It will be conformal invariant in the sense that

$$D^\chi(p(x,g)) \, \Gamma(g^{-1}x \, \chi \cdots \chi' g^{-1}x') \, D^{\chi'}(p(x',g)) = \Gamma(x\chi \cdots \chi' x').$$

(Dots... indicate dependence on further χ-variables). It must then vanish unless χ and χ' are equivalent, and then it is uniquely determined up to normalization by conformal invariance.

Because integrations $\oint d\chi$ could be taken over equivalence classes of unitary representations * we may without loss of generality put

$\Gamma(x\chi \ldots \chi' x') \propto 1\!\!1 \, \delta(\chi,\chi')\delta(x-x')$ with δ-function $\delta(\chi,\chi') = 0$ unless $\chi = \chi'$ and $\oint d\chi \, \delta(\chi,\chi') = 1$

In conclusion we find

---

* As it stands, every one of them is counted twice because $\chi \simeq \chi'$; but only the sum of contributions from equivalent representations is relevant.

Theorem 13.1. The vertex function admits an expansion of the following form, valid for all integers m in the range $1 \leq m \leq n$:

$$\Gamma(x_0 \ldots x_{n+1}) = \oint d\chi_1 \ldots \oint d\chi_{n-1} \sum_{s_2} \ldots \sum_{s_{n-1}} a_m(\{\chi\}, \{s\}) \cdot \tag{13.44a}$$
$$\int dx \; V(x_0 \ldots x_m ; x\chi_1\chi_2 s_2 \ldots \chi_m s_m)^* \, V(x_n \ldots x_{m+1} ; x\chi_{n-1}\chi_{n-2}s_{n-2} \ldots \chi_m s_m)$$

where $V(\cdots)$ resp. $V(\cdots)^*$ are column vectors resp. row vectors in $\mathcal{V}^{\ell_m}$ (for $\chi_m = [\ell_m, c_m]$) given in terms of Clebsch-Gordan kernels (theorem 10.1) by

$$V(x_0 \ldots x_m ; x\chi_1\chi_2 s_2 \ldots \chi_m s_m) = \int dy_1 \ldots dy_m \; V^{s_m}(x_m \tilde{\chi}_0 y_{m-1}\chi_{m-1} ; y_m \chi_m) \cdot \tag{13.44b}$$
$$\ldots V^{s_2}(x_2 \tilde{\chi}_0 y_1 \chi_1 ; y_2 \chi_2) V(x_0 \tilde{\chi}_0 x_1 \tilde{\chi}_0 ; y_1 \chi_1) .$$

Every factor herein is a matrix, except the last which is a vector in $\mathcal{V}^{\ell_1}$; the matrices are to be multiplied in the indicated order. $*$ denotes hermiteans conjugation of V . Summation over s_k ranges over UIR's of $\mathcal{U}$ contained both in ℓ_k and ℓ_{k-1} . ▮

All the dynamical information is in the complex function $a_m(\{\chi\}, \{s\})$ of $\chi_1 \cdots \chi_{n-1}$, and $s_2 \ldots s_{n-1}$. Factors $d(s_i) = (\dim s_i)^{-1}$ were absorbed into it.

By obvious modifications of the procedures which lead to Eqs. (13.39) and (13.44) many other expansion formulae can be obtained. For instance

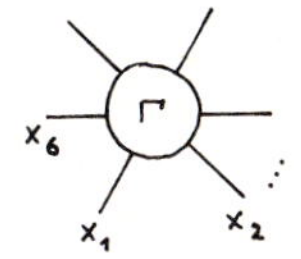

$$= \oint d\chi_1 \ldots \oint d\chi_3 \int V_\alpha(x_1 \tilde{\chi}_0 x_2 \tilde{\chi}_0 ; y_1 \tilde{\chi}_1) \, V_\beta(x_3 \tilde{\chi}_0 x_4 \tilde{\chi}_0 ; y_2 \tilde{\chi}_2) \tag{13.45}$$
$$\cdot V_\gamma(x_5 \tilde{\chi}_0 x_6 \tilde{\chi}_0 ; y_3 \tilde{\chi}_3) \Gamma_{\alpha\beta\gamma}(y_1\chi_1 \, y_2\chi_2 \, y_3\chi_3) \, dy_1 dy_2 dy_3$$

Indices α label a basis in $\mathcal{V}^{\ell_1}$, etc. The partial waves $\Gamma_{\ldots}(y_1\chi_1 \, y_2\chi_2 y_3\chi_3)$ herein are conformal invariant 3-point functions, and therefore linear combinations (with dynamically determined proportionality factors) of the CG-kernels of theorem 10.1.

In applications it is often convenient to use expansion formulae not for the (totally 1-particle irreducible) vertex functions, but for Green functions. They are obtained by first taking out the 1-particle reducible parts in those channels in which one wants to expand, and then proceding as above. We leave the details to the reader.

14. Implications of the dynamical equations. Pole structure of conformal partial waves

14.A. Poles in the conformal partial waves implied by the vertex bootstrap equations

We shall start with a brief review of the solution [M2] of the BS equation for the simple φ^3 model, and will then extend the results to the more realistic model of Sec. 12.D.

The 1PI amplitudes A_{1i} and the BS kernel B satisfy the same covariance and square integrability conditions (with respect to the arguments x_1, x_2) as the proper vertex functions Γ . We can therefore apply Eqs. (13.32), (13.34) to these functions:

$$A_{1i}(x_1x_2\,;\,x_3x_4) = \sum\!\!\!\!\!\!\int d\chi\; a(\chi)\, F_\chi(x_1x_2\,;\,x_3x_4) \tag{14.1}$$

$$B(x_1x_2\,;\,x_3x_4) = \sum\!\!\!\!\!\!\int d\chi\; b(\chi)\, F_\chi(x_1x_2\,;\,x_3x_4) \tag{14.2}$$

where

$$F_\chi(x_1x_2\,;\,x_3x_4) = \int dx\, V(x_1,-c_1\,;\,x_2,-c_2\,;\,x,\tilde{\chi})\, V(x_3,-c_3\,;\,x_4,-c_4\,;\,x,\chi) \tag{14.3}$$

and $a(\chi)$, $b(\chi)$ and F_χ also depend on the dimension parameters c_i of the fields *. Using the orthonormalization condition (11.9), we reduce the BS equation for the φ^3-model of Sec. 12 to the simple algebraic equation

$$a(\chi) = b(\chi) + b(\chi)\,a(\chi) \tag{14.4}$$

for the conformal partial waves. It implies that the partial wave amplitude

$$a(\chi) = \frac{b(\chi)}{1-b(\chi)} \tag{14.5}$$

has a pole for $\chi = \chi_\ell$ for which $b(\chi_\ell) = 1$.

Using the relation

$$a(\chi) = \gamma(\chi) + a_1(\chi) \tag{14.6}$$

where $a_1(\chi)$ is the partial wave of the sum of 1-particle reducible diagrams

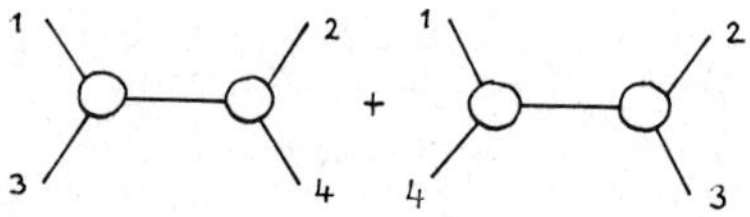

(14.7)

---

* The function $a(\chi)$ is related to the partial waves $g(\chi) = g(\chi; c_i)$ (i= 1,2,3,4) of the unamputated Green function G_{1i} used in ref. [M2] by $a(\chi) = g(\chi, -c_i)$. A similar relation holds for the corresponding BS kernels.

which has no singularities in the (12)-channel, we conclude that $\gamma(\chi)$ shares the poles of $a(\chi)$.

On the other hand, the bootstrap equation (12.14) for the vertex function (13.17) and the analytic continuation to real c of the orthonormality relation (11.9), imply

$$g\, V(x_1,-c\,;\, x_2,-c\,;\, x_3,-c) = g\, b(\chi_0)\, V(x_1,-c\,;\, x_2,-c\,;\, x_3,-c) \qquad \text{for } c = c_\varphi \tag{14.8}$$

A similar relation follows from Eq. (12.19). Thus,

$$b(\chi_0) = 1 \qquad \text{for} \quad \chi_0 = [0,-c] \qquad (c = c_\varphi) \tag{14.9a}$$

$$b(\chi_2) = 1 \qquad \text{for} \quad \chi_2 = [2,h] \tag{14.9b}$$

so that $a(\chi)$ and $\gamma(\chi)$ do have poles for $\chi = \chi_0$ and $\chi = \chi_2$.

Now we shall demonstrate that the same mechanism also works in the more complicated model of Sec. 12.D.

Let us consider the set of Green functions A_{1i} with total charge zero in the channel (1 2). We shall use the following shorthand notation for the corresponding partial waves:

$$\longleftrightarrow \quad a_{\varphi\varphi} \quad [= a(\varphi\varphi^* \to \varphi\varphi^*; \chi)\ ] \tag{14.10a}$$

$$\longleftrightarrow \quad a_{\varphi B} = a_{B\varphi} \ [= a(BB \to \varphi\varphi^*;\chi)] \tag{14.10b}$$

$$\longleftrightarrow \quad a_{BB} \quad [= a(BB \to BB;\chi)] \tag{14.10c}$$

and similarly for the BS amplitudes. The BS equations are reduced to a system of algebraic equations whose solution is given by

$$a_{\varphi\varphi} = \frac{1-b_{BB}}{\Delta} - 1 \ , \quad a_{B\varphi} = \frac{b_{B\varphi}}{\Delta} \ , \quad a_{BB} = \frac{1-b_{\varphi\varphi}}{\Delta} - 1 \tag{14.11}$$

$$\Delta = \Delta(\chi) = (1-b_{BB})(1-b_{\varphi\varphi}) - b_{B\varphi}^2 \tag{14.12}$$

On the other hand the bootstrap equations for the vertex functions (12.27) and (12.25) give

$$\begin{aligned} g_1 &= b_{B\varphi}(\chi_B)\, g_0 + b_{\varphi\varphi}(\chi_B)\, g_1 \\ g_0 &= b_{BB}(\chi_B)\, g_0 + b_{B\varphi}(\chi_B)\, g_1 \end{aligned} \tag{14.13}$$

The condition that the system (14.13) has a nontrivial solution with respect to the "coupling constants" g_o and g_1 leads to the equation

$$\Delta(\chi_B) = 0 \tag{14.14}$$

and thus imply the existence of a pole for $\chi = \chi_B$ of the amplitudes (14.11).

Similarly, starting from the equations for the 3-point functions (12.26), (12.27), which involve the stress energy tensor, we obtain

$$\Delta(\chi_2) = 0 \tag{14.15}$$

for χ_2 given by (14.9b). Finally, the bootstrap equation for the current-field 3-point function (12.28) and the vanishing of $\langle T\mathcal{B}(x_1)\mathcal{B}(x_2)j_\mu(x_3)\rangle_0$ give

$$b_{B\varphi}(\chi_1) = 0 \quad , \quad b_{\varphi\varphi}(\chi_1) = 1 \quad \Rightarrow \quad \Delta(\chi_1) = 0 \tag{14.16a}$$

for

$$\chi_1 = [1, h-1] . \tag{14.16b}$$

14.B. Pole structure of the n-point partial waves. Expression for the residues

In this subsection we shall spell out the implications of the dynamical equations for the simplest (φ^3) model only. The extension of the results to the (φ, φ^*, B) model of Sec. 12.D which uses (14.11-14.16),is quite straightforward.

First of all we shall demonstrate that the poles of $\gamma(\chi)$ (and $a(\chi)$, corresponding to the points (14.9), are also poles of the n-point partial waves $\Gamma_\chi(x; x_3 \dots x_n)$ (13.33) for all $n \geqslant 4$. We deduce this statement in two steps. It is true, if we replace Γ_χ by the $1i$-partial wave

$$A^{\chi}_{1i}(x; x_3 \cdots x_n) = \frac{1}{2}\int dx_1 \int dx_2\, V(x_1, c_\varphi; x_2, c_\varphi; x\chi)\, A_{1i}(x_1 x_2; x_3 \cdots x_n) \tag{14.17}$$

Indeed,taking the conformal Fourier transform of Eq. (12.12) we obtain

$$[1-b(\chi)]\, A^{\chi}_{1i}(x; x_3 \cdots x_n) = A^{\chi}_{2i}(x; x_3 \cdots x_n) + b(\chi)\sum\int dy_1 \int dy_2\, V(y_1 c_\varphi; y_2 c_\varphi; x\chi) A(y_1 x_3 \dots) A(y_2 \dots x_{i_n})$$
$$+ \sum_{k=3}^{n} \sum \int \cdots \int dy_1 \dots dy_k\, A^{\chi}_{2i}(x; y_1 \dots y_k)\, G(y_1 - y_1') \dots G(y_k - y_k')\, A(y_1' x_3 \dots) \cdots A(y_k' x_{i_k} \cdots) \tag{14.18}$$

It follows that for each $\chi = \chi_\ell$ for which

$$b(\chi_\ell) = 1 \tag{14.19}$$

(and the right hand side of (14.18) does not vanish) the partial wave A^{χ}_{1i}

must have a pole. This is true, in particular, for $\chi = \chi_0$ and $\chi = \chi_2$, because of (14.9). It remains to show (as a second step) that the conformal Fourier transforms Γ_χ of the proper vertex functions also have poles in these points. That follows from the observation that the difference between A_{1i}^{χ} and Γ_χ is given by a convergent skeleton diagram, provided that c_φ satisfies (13.23) (see Sec. 13.C and ref. [D2]).

We shall assume at this point that we are dealing with simple poles only, so that

$$b_\ell = \left.\frac{db(\chi)}{dc}\right|_{\chi=\chi_\ell} \neq 0 \tag{14.20}$$

The conformal expansion of the 2i-kernel in the dynamical equations (12.13) leads to the relation

$$\Gamma(x; x_3 \ldots x_n) = g\, A_{2i}^{\chi_0}(x; x_3 \ldots x_n) \tag{14.21}$$

Noting the relation between the amputated Green function A and the proper vertex function Γ and combining (14.18), (14.9a), (14.20), (12.12) and (14.21) we can express the residue of A_{1i} (or Γ_χ) at the pole $\chi = \chi_0$ in terms of the amputated Green function A:

$$-g b_0 \operatorname*{Res}_{\chi=\chi_0} A_{1i}^{\chi}(x; x_3 \ldots x_n) = -g b_0 \operatorname*{Res}_{\chi=\chi_0} \Gamma_\chi(x, x_3, \ldots x_n) = A(x, x_3, \ldots, x_n) \tag{14.22}$$

Similarly, the residue of A_{1i}^{χ} (or Γ_χ) at $\chi = \chi_2$ can be expressed in terms of the amputated (n-1)-point function $A_{\mu\nu}(x; x_3 \ldots x_n)$, which involves the stress energy tensor $\Theta_{\mu\nu}(x)$.

14.C. Basic conformal covariant tensor fields. Analyticity assumption

The preceeding argument can be generalized as follows. Let $O_\ell(x,\mathfrak{z})$ be a conformal covariant rank ℓ tensor-field of dimension $h + c_\ell$ for which the 3-point function

$$\langle T \varphi(x_1)\, \varphi^*(x_2)\, O_\ell(x,\mathfrak{z}) \rangle_0 \Leftrightarrow G_\ell(x_1, x_2; x\mathfrak{z}) \tag{14.23}$$

does not vanish. (In writing $O_\ell(x,\mathfrak{z})$ in Minkowski space, we can regard

ζ as a real light vector,--cf. [T6]). Then the conformal partial waves $\Gamma_\chi(x; x_3 \ldots x_n)$ have a pole for $\chi = \chi_\ell = [\ell, c_\ell]$. The argument is the same as before.

Let $\varphi(x)$ be a free 0-mass field; in other words let $\varphi(x)$ satisfy the D'Alambert equation

$$\nabla^2 \varphi(x) \equiv \Box \varphi(x) = 0 \tag{14.24}$$

and the canonical commutation relations. Consider the bi-local operator

$$O_\ell(x_1, x_2; \zeta) = \kappa_\ell i^\ell \mathcal{D}_\ell(h-1, \zeta\nabla_1; h-1, \zeta\nabla_2)\, \varphi^*(x_1)\, \varphi(x_2) \tag{14.25a}$$

where D_ℓ (a, α; b, β) is the polynomial (11.5), and κ_ℓ is a normalization constant $(\zeta\nabla = \zeta^0\nabla^0 - \underline{\zeta}\underline{\nabla})$. In the simple case at hand (in which a = b = h - 1) D_ℓ can be expressed in terms of a Gegenbauer polynomial:

$$\mathcal{D}_\ell(h-1, \alpha; h-1, \beta) = \ell!\,(\alpha+\beta)^\ell\, P_\ell^{(h-2,h-2)}\left(\frac{\alpha-\beta}{\alpha+\beta}\right) = \ell!\,\frac{(h-1)_\ell}{(2h-3)_\ell}\,(\alpha+\beta)^\ell\, C_\ell^{h-3/2}\left(\frac{\alpha-\beta}{\alpha+\beta}\right) \tag{14.25b}$$

(see, e.g., [G7] Eqs. 8.932 and 8.962.4; we have again used the short-hand notation $(a)_\ell$ of (5.11)). We note that for the physically interesting case of 4-dimensional space-time h-1 = 2h-3 = 1 and $C_\ell^{1/2}$ coincides with the Legendre polynomial. The relevant property of the polynomials $\mathcal{D}_\ell$ for our present purposes is given by the following differential relations, valid for $\nabla_1^2 = 0 = \nabla_2^2$ which can be assumed because of (14.24) :

$$\begin{aligned}(\nabla_1 \cdot D)\mathcal{D}_\ell(h-1, \zeta\nabla_1; h-1, \zeta\nabla_2) &= \ell(\ell+h-2)^2 \mathcal{D}_{\ell-1}(h-1, \zeta\nabla_1; h-1, \zeta\nabla_2)\, \nabla_1\nabla_2 \\ &= -(\nabla_2 \cdot D)\mathcal{D}_\ell(h-1, \zeta\nabla_1; h-1, \zeta\nabla_2)\end{aligned} \tag{14.26}$$

where D is the interior differentiation (A.47), (A.49) on the light cone. Eq. (14.26) implies that the local operator

$$O_\ell(x, \zeta) = :O_\ell(x, x; \zeta): = \lim_{x_1, x_2 \to x} [O_\ell(x_1, x_2; \zeta) - \langle O_\ell(x_1, x_2; \zeta)\rangle_0] \tag{14.27}$$

is a conserved tensor-current:

$$(D \cdot \nabla)\, O_\ell(x, \zeta) = 0 \tag{14.28}$$

Moreover, it is a (weakly [H6]) conformal covariant basic tensor (in the sense of ref.[T4]). We recall that for a basic tensor $O_\ell(x)$ the infinitesimal generators of special conformal transformation vanish at x=0. This means that the Euclidean counterpart of O_ℓ transforms under an elementary representation of type χ_ℓ of $O^\uparrow(2h+1,1)$. In fact, for any vector $|\Psi\rangle$ with a finite energy the function $F_\ell(x,\mathfrak{z}) = \langle\Psi| O_\ell(x,\mathfrak{z})|0\rangle$ can be continued analytically into the extended (complex) tube (see[B5]); its restriction to Euclidean x transforms according to the irreducible part of the "canonical representation"

$$\chi_\ell^{can} = [\ell, h+\ell-2] \tag{14.29}$$

(It belongs to the space $\mathcal{D}_{\ell-1,1}$ --cf. Sec. 6. We caution the reader that the gradient of a basic tensor is in general not a basic tensor-- cf. [T4]). If $\varphi(x)$ is a complex field ($\varphi \neq \varphi^*$) then the 3-point function (14.23) does not vanish for any of the operators (14.27). Choosing the normalization constant $\kappa_1 = e/h-1$, where e is the charge carried by φ^*, we can identify O_1 with the electromagnetic current:

$$O_1(x,\mathfrak{z}) = \mathfrak{z}^\mu j_\mu(x) \ ; \quad j_\mu(x) = ie\left[\varphi^*(x)(\nabla_\mu\varphi(x)) - \varphi(x)(\nabla_\mu\varphi^*(x))\right] \tag{14.30}$$

If φ is a neutral field ($\varphi = \varphi^*$), then the fields $O_\ell(x,\mathfrak{z})$ with ℓ odd vanish. Setting $2h(2h-1)\kappa_2 = 1$ we obtain as a special case the (traceless) stress-energy tensor for $\ell = 2$:

$$O_2(x,\mathfrak{z}) = \mathfrak{z}^\mu\mathfrak{z}^\nu\Theta_{\mu\nu}(x) \tag{14.31a}$$

$$\Theta_{\mu\nu}(x) = :\nabla_\mu\varphi(x)\nabla_\nu\varphi(x): - \tfrac{1}{2}g_{\mu\nu}:\nabla^\lambda\varphi(x)\nabla_\lambda\varphi(x): + \\ + \frac{1}{2}\frac{h-1}{2h-1}(\Box g_{\mu\nu} - \nabla_\mu\nabla_\nu):\varphi^2(x): \tag{14.31b}$$

We shall take the case of a free field as a guide concerning the set of basic tensor fields coupled to φ in general. We shall assume, in particular, in the case of the (neutral) φ^3 model of Sec. 12, that for each even ℓ there exists at least one basic "composite" field for which the 3-point function (14.23) with φ does not vanish. There is no reason to believe that for $\ell \geq 4$ the dimension of the field -- in a non-trivial, interacting theory-- is canonical (i.e., that χ_ℓ is given by (14.29)). However, positivity of the 2-point Wightman function of O_ℓ implies that if $\chi_\ell = [\ell, c_\ell]$ is the $O^\uparrow(2h+1,1)$-representation label for O_ℓ, then

$$c_\ell \geq h+\ell-2\,. \tag{14.32}$$

(see [R2, F2] and Sec. 5.D).

Thus, the dynamical equations and our assumption about the set of composite basic fields imply the existence of a denumerable infinity of poles $\chi \equiv \chi_\ell$ in the conformal partial waves, satisfying (14.32). It is natural to conjecture that these are the only singularities of Γ_χ in the right half plane c . More precisely, we shall postulate that $\gamma(\chi)$ and $\Gamma_\chi(x;x_3\ldots x_n)$ are meromorphic functions of c in the right half plane with simple poles, restricted to the real c axis. We remark that unlike a similarly sounding ansatz about the singularity structure of scattering amplitudes in the complex angular momentum plane, this postulate is not in conflict with (off-shell) unitarity, since the dynamical equations are taken exactly into account.

15. Derivation of an operator product expansion for vacuum expectation values

15.A. Another form of the conformal expansion, involving a Minkowski momentum space integral. The Q-kernels.

In order to exploit the postulate about the meromorphic structure of Γ_χ it would be natural to try to close the integration path in the representation (13.32) in the right half plane c and then apply the residue theorem. In doing that, however, one encounters the problem of the asymptotic behavior in c of the integrand. A straightforward way to analyze it consists in performing a partial Fourier transform of the vertex functions. To this end we replace the x-space integration in (13.32) by

$$\Gamma(x_1\ldots x_n) = \not{\Sigma}\, d\chi \int (dp)\, V_-^{\tilde{\chi}}(x_1 x_2; -p)\, G_\chi(p)\, \Gamma_{\tilde{\chi}}(p; x_3\ldots x_n)\ , \quad (dp) = \frac{d^{2h}p}{(2\pi)^h} \tag{15.1}$$

where $G_\chi(p)$ is the 2-point Green function (13.13) and

$$\begin{aligned} V_\pm^{\tilde{\chi}}(x_1 x_2; -p) &= \int V(x_1, \pm c_1; x_2, \pm c_2; x_3, \tilde{\chi})\, e^{ipx_3}\, dx_3 \\ &= 2A_\ell^\pm \left(\frac{2}{x_{12}^2}\right)^{h-\delta_{\tilde{\chi}} \pm c_+} \mathcal{D}_\ell(\delta_{\tilde{\chi}} \pm c_-, 3\nabla_1; \delta_{\tilde{\chi}} \mp c_-, 3\nabla_2)\left(\frac{x_{12}^2}{p^2}\right)^{\frac{\ell+c}{2}} \\ &\quad \int_0^1 du\, f_\pm(u)\, e^{ip(ux_{12}+x_2)}\, K_{\ell+c}\left(\sqrt{u(1-u)x_{12}^2 p^2}\right) \end{aligned} \tag{15.2}$$

$$f_\pm(u) = [u(1-u)]^{\frac{h}{2}-1}\left(\frac{u}{1-u}\right)^{\pm c_-}$$

where D is the ℓ-th order differential operator (11.5) and $A_\ell^{\pm}$ is given by Eq. (F.4) of Appendix F.1, where the last equation is derived. For large ℓ it is not obvious that the integral in u in (15.2) makes sense. In order to write (15.1) in a manifestly meaningful form, which will be at the same time convenient for studying the asymptotic properties of conformal partial waves in c, we shall assume that

$$\sigma_1 < 0,\ \sigma_2 < 0;\ \sigma_3 > 0, \dots, \sigma_n > 0 \quad , \text{ where } \sigma_j = (x_j)_{2h}, \tag{15.3}$$
$$x_j \neq x_k \text{ for } j \neq k ,$$

and shall shift the integration path in the complex p_{2h}-plane to a contour C around the negative imaginary semi-axis (see Fig.3). In the domain (15.3) the exponential in the right hand side of (15.2) decreases for $p_{2h} \to -i\infty$; so does also $\Gamma_{\tilde{\chi}}(p;x_3,\dots x_n)$ because of the spectral condition. Hence, the shift of the complex energy integration path indicated above is indeed legitimate, and we are led to evaluate the discontinuity

$$i\Delta = \operatorname{disc}\left[V_-^{\tilde{\chi}}(x_1,x_2;-p)\, G_\chi(p)\, \Gamma_{\tilde{\chi}}(p;x_3,\dots,x_n)\right], \tag{15.4a}$$

$$\operatorname{disc} f(p) = \theta(p_0) \lim_{\varepsilon \downarrow +0} \left[f(\underline{p}, -ip_0+\varepsilon) - f(\underline{p}, -ip_0-\varepsilon)\right]. \tag{15.4b}$$

We can directly compute (15.4) in the case n=4 using the explicit expression (15.2) for the vertex function. To handle the general case we shall assume that the partial wave for all $n \geqslant 4$ has the form

$$\Gamma_{\tilde{\chi}}(p;x_3,\dots,x_n) = f_1^{\tilde{\chi}}(p;x_3,\dots,x_n) + G_{\tilde{\chi}}(p) f_2^{\chi}(p;x_3,\dots,x_n) \tag{15.5}$$

$$\left(= \left(\frac{p^2}{2}\right)^{-c-\ell} f_2'^{\chi}(p;x_3,\dots,x_n) + f_1^{\tilde{\chi}}(p;x_3,\dots,x_n)\right),$$

where the functions $f_1^{\tilde{\chi}}(p)$ and $f_2^{(\prime)\chi}(p)$ satisfy

$$disc\, f_1^{\tilde{\chi}}(p) = 0 = disc\, f_2^{\prime\prime\chi}(p).$$

This assumption can be justified in the framework of the skeleton perturbation theory by writing

$$\Gamma^{\tilde{\chi}}(p; x_3, \ldots, x_n) = \iint dx' dx'' \, V_-^{\tilde{\chi}}(x', x''; p)\, S(x', x''; x_3, \ldots, x_n),$$

where $S(x', x'', x_3, \ldots x_n)$ is a sum of skeleton diagrams. To this end it suffices to express K_ν from the identity

$$2\sin\pi\nu\, K_\nu(z) = \pi\left[I_{-\nu}(z) - I_\nu(z)\right],$$

(see, e.g., Sec. 8.4 of ref. [G7]) in the representation (15.2) for $V_-^{\tilde{\chi}}$ that gives

$$V_\pm^{\tilde{\chi}}(x_1, x_2; -p) = \frac{\pi}{\sin\pi(l+c)}\left[Q_{1\pm}^{\tilde{\chi}}(x_1, x_2; -p) - G_{\tilde{\chi}}(p)\, Q_{2\pm}^{\chi}(x_1, x_2; p)\right], \quad (15.6)$$

where $Q_1(G_{\tilde{\chi}}Q_2)$ is obtained from (15.2) by replacing $2K_{l+c}$ with $-I_{l+c}$ $(-I_{-l-c})$.

For $p_{2h} \to -ip_0 \pm 0$ we have $p^2 \to -p_M^2 \mp i0p_0$ where $p_M^2 = p_0^2 - \underline{p}^2$, $p_M \equiv (p_0, \underline{p})$, $p = (ip_{2h}, \underline{p})$. To evaluate $\mathrm{disc}\, G_\chi$ we use the relations

$$(-p_M^2 \pm i0)^c = (-p^2)_+^c + e^{\pm i\pi c}(p_M^2)_+^c \quad , \quad (x_+^c \equiv \theta(x)|x|^c), \quad (15.7a)$$

$$\Pi^{ls}(p, \zeta, \zeta') \longrightarrow (-1)^l\, \Pi^{ls}(p_M; \zeta_M, \zeta'_M), \quad (15.7b)$$

(for $p_{2h} \to -ip_0$)

$$[\zeta_M = (\zeta_0, \underline{\zeta}),\ \zeta_0 = i\zeta_{2h}].$$

From (13.13) and (15.7) we obtain

$$disc\, G_\chi(p) = 2i\sin\pi(l-c)\, w_\chi(p_M), \quad (15.8)$$

where

$$W_{\chi}(p) = \theta_+(p)\left(\tfrac{1}{2}p^2\right)^c \sum_{s=0}^{l} \alpha_s(c) \Pi^{l_s}(p), \quad p = p_M, \quad \theta_+(p) = \theta(p_0)\theta(p^2) \tag{15.9}$$

is the Wightman function, normalized by

$$W_{\chi}(p) W_{\tilde{\chi}}(p) = \theta_+(p)\mathbb{1}, \quad W_{\chi}(p) W_{\tilde{\chi}}(p) W_{\chi}(p) = W_{\chi}(p) \tag{15.9}$$

Now we are ready to evaluate Δ in (15.4); using (15.5), (15.8) and (4.11), we obtain

$$\Delta = 2\pi \Big\{ Q_{2-}^{\chi}(x_1, x_2; -\underline{p}, ip_0) W_{\tilde{\chi}}(p_M) f_2^{\chi}(\underline{p}, -ip_0; x_3, \ldots, x_n) - \\ - Q_{1-}^{\tilde{\chi}}(x_1, x_2; -\underline{p}, ip_0) W_{\chi}(p_M) f_1^{\tilde{\chi}}(\underline{p}, -ip_0; x_3, \ldots, x_n) \Big\}. \tag{15.10}$$

In order to relate the functions $f_1^{\tilde{\chi}}$ and f_2^{χ}, and Q_1 and Q_2 between themselves we shall use the symmetry property (13.37). First we notice that, according to (15.5),

$$\mathrm{disc}\, \Gamma_{\tilde{\chi}}(p; x_3, \ldots, x_n) = 2i \sin\pi(l+c) W_{\tilde{\chi}}(p_M) f_2^{\chi}(\underline{p}, -ip_0; x_3, \ldots, x_n). \tag{15.11a}$$

Taking into account (13.37) we find on the other hand

$$\mathrm{disc}\, \Gamma_{\chi}(p; x_3, \ldots, x_n) = \mathrm{disc}\, G_{\chi}(p) \Gamma_{\tilde{\chi}}(p; x_3, \ldots, x_n) = \\ = -2i \sin(l+c) W_{\chi}(p_M) f_1^{\tilde{\chi}}(\underline{p}, -ip_0; x_3, \ldots, x_n). \tag{15.11b}$$

It follows that

$$\theta_+(p_M) f_1^{\tilde{\chi}}(\underline{p}, -ip_0; x_3, \ldots, x_n) = \theta_+(p_M) f_2^{\tilde{\chi}}(\underline{p}, -ip_0; x_3, \ldots, x_n) = \\ = \frac{1}{2i \sin\pi(l-c)} W_{\tilde{\chi}}(p_M)\, \mathrm{disc}\, \Gamma_{\chi}(p; x_3, \ldots, x_n), \tag{15.12a}$$

and, in particular,

$$\theta_+(p_M) Q_{1\pm}^{\tilde{\chi}}(x_1, x_2; \underline{p}, -ip_0) = \theta_+(p_M) Q_{2\pm}^{\tilde{\chi}}(x_1, x_2; \underline{p}, -ip_0) = \\ = \frac{1}{2\pi i} W_{\tilde{\chi}}(p_M)\, \mathrm{disc}\, V^{\chi}(x_1, x_2; p) = \theta_+(p_M) Q_{\pm}^{\tilde{\chi}}(x_1, x_2; p_M), \tag{15.12b}$$

where we shall use the following expression for $Q_{\pm}^{\tilde{\chi}}$:

$$Q_{\pm}^{\tilde{\chi}}(x_1, x_2; P_M) \equiv - A_\ell^{\pm} \left(\frac{2}{x_{12}^2}\right)^{h - \delta_{\tilde{\chi}} \pm c_+} D_\ell(\delta_{\tilde{\chi}} \pm c_-, \mathfrak{z}\nabla_1; \delta_{\tilde{\chi}} \mp c_-, \mathfrak{z}\nabla_2) \left(\frac{x_{12}^2}{P_M^2}\right)^{\frac{\ell+c}{2}} \cdot$$

$$\cdot \int_0^1 f_{\pm}(u) e^{-iP_M(u x_{12} + x_2)} J_{\ell+c}\left(\sqrt{u(1-u) x_{12}^2 P_M^2}\right) du, \tag{15.12c}$$

and the scalar product in the exponent is $xP_M = \underline{x}\underline{P} - i x_{2h} P_0$.

Eqs. (15.11) and (15.6), (15.12) suggest to define the functions $Q_{\Gamma}^{\tilde{\chi}}(p\ ;\ x_3, \ldots x_n)$ by

$$W_\chi(P_M)\, Q_{\Gamma}^{\tilde{\chi}}(P_M; x_3, \ldots, x_n) = \frac{1}{2\pi i}\, \mathrm{disc}\, \Gamma_\chi(P; x_3, \ldots, x_n), \tag{15.13a}$$

or equivalently (due to (15.9′))

$$\theta_+(P_M)\, Q_{\Gamma}^{\tilde{\chi}}(P_M; x_3, \ldots, x_n) = \frac{1}{2\pi i}\, W_{\tilde{\chi}}(P_M)\, \mathrm{disc}\, \Gamma_\chi(P; x_3, \ldots, x_n). \tag{15.13b}$$

Eqs. (15.10), (15.12) in the above notation and the symmetry of the range of c-integration in (15.1) allow us to derive the following representation for the proper vertex function

$$\Gamma(x_1, \ldots, x_n) = \oint d\chi \int (dP_M)\, \Delta \ =$$

$$= -(2\pi)^2 \oint \frac{d\chi}{\sin\pi(\ell+c)} \int (dP_M)\, Q_-^{\tilde{\chi}}(x_1, x_2; -P_M)\, W_\chi(P_M)\, Q_{\Gamma}^{\tilde{\chi}}(P_M; x_3, \ldots, x_n). \tag{15.14}$$

In the special case n=4, using (13.34) we can rewrite (15.14) in the form

$$\Gamma(x_1, x_2, x_3, x_4) = -(2\pi)^2 \oint d\chi \frac{\gamma(\chi)}{\sin\pi(\ell+c)} \int (dP_M)\, Q_-^{\tilde{\chi}}(x_1, x_2; -P_M)\, W_\chi(P_M)\, Q_-^{\tilde{\chi}}(x_3, x_4; P_M). \tag{15.14′}$$

Unlike Eq. (15.2) which, as noted, involves a divergent integral (in u) for $\ell \geq h/2$ (and any c) the integral in the right hand side of (15.12), which defines the functions $Q_{\pm}^{\tilde{\chi}}$, is absolutely convergent for

$$\ell + h + \mathrm{Re}\, c > 2|c_-|. \tag{15.15}$$

Moreover, the representation (15.14′) allows to exploit the analytic structure of $\gamma(\chi)$ by deforming the integration path in the right half plane c. That possibility is secured by the following characteristic properties of $Q_{\pm}^{\tilde{\chi}}$.

(i) $Q_{\pm}^{\tilde{\chi}}(x_1 x_2 ; -p 3)$ is an entire analytic funtion of p (provided that the integral in u converges, which is certainly true in the range (15.15)).

(ii) For time-like vectors p, $Q^{\tilde{\chi}}$ decreases exponentially for Re $c \to \infty$. (That property, which will enable us to close the integration path in c, follows from the known asymptotic behavior of the Bessel function J (x) for $\nu \to \infty$, - see, e.g. [G7] Eq. 8.452.1).

(iii) For small x_{12}, $Q_{-}^{\tilde{\chi}}$ is given by

$$Q_{-}^{\tilde{\chi}}(x_1 x_2 ; -p \cdot 3) \underset{x_{12}\to 0}{\approx} N_\ell(-c_+, -c_-, -c) \frac{\Gamma(c)\ (h-c-1)_\ell}{\Gamma(-c-\ell)\Gamma(h+c+\ell)} e^{ipx_2} \left(\frac{2}{x_{12}^2}\right)^{h-\delta_\chi-c_+} (3x_{12})^\ell \qquad (15.16)$$

It is natural to assume that in the general case $n > 4$ the kernel Q_r, given by (15.13) satisfies the properties (i) and (ii). This assumption can be justified again in the framework of the skeleton perturbation theory.

15.B. The vacuum operator product expansion

Now it is legitimate to close the contour of c integration in (15.14) in the right half plane, assuming the partial waves are sufficiently well behaved at infinity. However, transforming the integral over $V^{\tilde{\chi}}$ into an integral over $Q^{\tilde{\chi}}$ (which vanishes for Re $c \to \infty$) we have paid a certain price: the appearance of the factor $[\sin \pi (\ell + c)]^{-1}$ which introduces new "kinematical" poles. The main purpose of this subsection is to demonstrate that these poles are actually cancelled out.

First of all, we note that a finite number of poles coming from the sine factor are cancelled by the zeros of the Plancherel weight (11.11) for

$$c = 0, \ldots, h-2, \ h+\ell-1 \qquad (15.17)$$

(At this point we assume, for the sake of simplicity, that h is a positive integer, which includes the cases h=2 and h=3 we are primarily interested in. The argument--and the result--can also be extended to the case of half odd integer h). There remain two (infinite) sequences of "kinematical" poles to be dealt with; they correspond to the elementary representations $[\ell, c]$ with labels

$$c = h-1+\ell+\nu \quad , \quad \ell = 0,1,2,\dots \quad , \quad \nu = 1,2,\dots \tag{15.18a}$$

and

$$c' = h-1+\ell \quad , \quad \ell' = \ell+\nu \quad (\ell = 0,1,2,\dots, \nu = 1,2,\dots) \tag{15.18b}$$

which satisfy

$$c'+\ell' = c+\ell = h-1+2\ell+\nu \tag{15.19}$$

The clue to the cancellation problem lies in the partial equivalence of the representations

$$\chi^{+}_{\ell\nu} = [\ell, c] \;\; (= [\ell, h-1+\ell+\nu]\,) \quad \text{and} \quad \chi'^{+}_{\ell\nu} = [\ell', c'] \;\; (= [\ell+\nu, h-1+\ell]\,) \tag{15.20}$$

exhibited in Sec. 6.B. It leads, in particular, to the following identities among Q-kernels and conformal partial waves (see Appendix G):

$$(ip\mathfrak{z})^{\nu}\, Q_{-}^{\chi^{-}_{\ell\nu}}(x_1,x_2\,;-p,\mathfrak{z}) = \operatorname{sign}\left[\left(\tfrac{1-\nu}{2}+c_{-}\right)_{\nu}\right] Q_{-}^{\chi'^{-}_{\ell\nu}}(x_1,x_2\,;-p,\mathfrak{z})\,, \tag{15.21}$$

$$(-ip\mathfrak{z})^{\nu}\, Q_{r}^{\chi^{-}_{\ell\nu}}(p,\mathfrak{z}\,;x_3\dots x_n) = \operatorname{sign}\left[\left(\tfrac{1-\nu}{2}-c_{-}\right)_{\nu}\right] Q_{r}^{\chi'^{-}_{\ell\nu}}(p,\mathfrak{z}\,;x_3,\dots,x_n) \tag{15.22}$$

where

$$\chi^{(\prime)-}_{\ell\nu} = \tilde{\chi}^{(\prime)+}_{\ell\nu} = [\ell^{(\prime)}, -c^{(\prime)}]\,. \tag{15.20'}$$

Furthermore, we shall use the relation

$$[(\ell+1)_{\nu}]^{-2}(p\cdot\partial_{\mathfrak{z}})^{\nu}\, w_{\chi'^{+}_{\ell\nu}}(p\,;\mathfrak{z},\partial_{\mathfrak{z}})\,(p\mathfrak{z})^{\nu} = (-1)^{\nu}(h+\ell-1)_{\nu}^{-1}(2h+\ell-2)_{\nu}\, w_{\chi^{+}_{\ell\nu}}(p\,;\mathfrak{z},\partial_{\mathfrak{z}}) \tag{15.23}$$

which follows from (6.32)(15.9) and the identity

$$(\rho(\chi'^{+}_{\ell\nu}) \equiv)\;\; \rho_{\ell+\nu}(h-1+\ell) = -\frac{(h+\ell-1)_{\nu}}{(2h+\ell-2)_{\nu}}\, \rho_{\ell}(h-1+\ell+\nu)\,, \tag{15.24}$$

which is a direct consequence of the definition (11.11) of the Plancherel weight.

Now we are ready to prove that the sum of the residue s in the kinematical poles $\chi^{+}_{\ell\nu}$ and $\chi'^{+}_{\ell\nu}$ vanishes.

Indeed, due to (15.19), for both these poles

$$\pi \operatorname{Res} [\sin \pi (\ell + c)]^{-1} = \pi \operatorname{Res} [\sin \pi (\ell' + c')]^{-1} = (-1)^{h-1+\nu} \tag{15.25}$$

and we have (according to (15.21)-(15.24))

$$[(\ell+\nu)!]^{-2} \rho(\chi'^{+}_{\ell\nu}) Q_{-}^{\chi'^{-}_{\ell\nu}}(x_1, x_2; -p, \partial_\zeta)\, w_{\chi'^{+}_{\ell\nu}}(p; \zeta, \partial_3)\, Q_{\Gamma}^{\chi'^{-}_{\ell\nu}}(p, 3; x_3, \dots, x_n) =$$

$$= -[\ell!]^{-2} \rho(\chi^{+}_{\ell\nu}) Q_{-}^{\chi^{-}_{\ell\nu}}(x_1, x_2; -p, \partial_\zeta)\, w_{\chi^{+}_{\ell\nu}}(p; \zeta, \partial_3)\, Q_{\Gamma}^{\chi^{-}_{\ell\nu}}(p, 3; x_3, \dots, x_n). \tag{15.26}$$

This proves the cancellation of the kinematical poles coming from $[\sin \pi\ (\ell+c)]^{-1}$.

Let us note that in (15.26), $w_{\chi'^{+}_{\ell\nu}}(p)$ is proportional to the Minkowski space analogue of the kernel (6.28), which defines the hermitean form (6.26). This function is given by a formula similar to (6.23'):

$$w_{\chi'^{+}_{\ell\nu}}(p) = \lim_{\varepsilon \to 0} \left[\frac{d}{d\varepsilon} (\varepsilon\, w_{\chi'^{+}_{\varepsilon}}) \right. \tag{6.23''}$$

Taking into account that the singularity of the integrand for $\chi = \chi'^{+}_{\ell\nu}$ is in fact a double pole, we are led to use Eq. (6.23'') which yields the left hand side of (15.26).

Thus, closing the contour of integration in (15.14) in the right half-plane c we obtain a representation of the proper vertex function as a sum over dynamical poles and poles coming from the normalization factor of $Q_{-}^{\tilde{\chi}}$ only:

$$\Gamma(x_1, \dots, x_n) =$$
$$= 2\pi \sum_{\chi_\ell} \operatorname*{Res}_{\chi = \chi_\ell} \left\{ \frac{\rho_\ell(c)}{\sin \pi(\ell+c)} \int (dp)\, Q_{-}^{\tilde{\chi}}(x_1, x_2; -p)\, w_\chi(p)\, Q_{\Gamma}^{\tilde{\chi}}(p; x_3 \dots x_n) \right\} \tag{15.27}$$

A similar expansion can be deduced from here for the full Green function (see subsection 15.C below). It can be regarded as the result of inserting a conformal covariant operator product expansion (of the type considered in refs. [B6, G9, F1, F3, F4, F5, M5, S3]) in the Wightman functions (which is subsequently continued analytically in the Euclidean region). It is, however, important for our derivation that the operator product $\varphi(x_1)\, \varphi(x_2)$ (which is effectively decomposed) acts directly on the vacuum. (That was used in exploiting the inequalities (15.3) for the Euclidean time-components). It is indicated in ref. [S4] that the general (global) operator product expansion is more complicated. That is why we adopt the term "vacuum expansion" (of ref. [S4]) for the situation envisaged here.

<u>Remark.</u> It follows from (15.21), (13.34) and (15.22) (for n = 4) that

$$\left(\frac{1-\nu}{2}+c_-\right)_\nu \gamma(\chi^+_{\ell\nu}) = \left(\frac{1-\nu}{2}-c_-\right)_\nu \gamma(\chi'^+_{\ell\nu}) \tag{15.28a}$$

as a consequence, (according to Eq.[G3])

$$\gamma(\chi^+_{\ell\nu}) = (-1)^\nu \gamma(\chi'^+_{\ell\nu}) \quad \text{for } c_- \neq 0 . \tag{15.28b}$$

On the other hand, if $c_- = 0$ and $\varphi_1(x) = \varphi_2(x) = \varphi(x)$, then $\gamma(\chi_\ell) = 0$ for odd ℓ, and if $\gamma(\chi_\ell) \neq 0$ for even ℓ, Eq. (15.28b) cannot hold for odd ν.

Similarly, the limit of the Clebsch-Gordan kernel V for $c_- \to 0$ does not commute with the one for $\chi \to \chi^+_{\ell\nu}$. This complicates the treatment of the points $\chi'^+_{\ell\nu}, \chi^+_{\ell\nu}$ in which the partial waves $Q_r^{\tilde{\chi}}(p,3;x_3x_4)$ also have a pole (for $c_- = 0$). The result can also be extended to that case by an appropriate cancellation of the higher order poles of the integrand.

15.C. Wightman positivity for the 4-point function

The representation (15.27) is particularly convenient in analyzing the positivity properties of the 4-point Wightman function (cf. [M3, O2]). It is, of course, the full 4-point (Wightman or Schwinger) function that exhibits positivity, and not just the proper vertex function Γ. So, our first task will be to write down the counterpart of (15.27) for the Schwinger function

$$s(x_1, x_2, x_3, x_4) = \langle \phi_1(x_1) \phi_2(x_2) \phi_1(x_3) \phi_2(x_4) \rangle_0$$

where ϕ_1 and ϕ_2 are spinless Euclidean fields with dimension parameters c_1 and c_2 (we can have $\phi_1 = \phi_2$ as a special case).

Let us assume that there is a scalar (Euclidean) field $\phi_3(x)$ with $c_3 < 0$ such that the 3-point function $\langle \phi_1(x_1) \phi_2(x_2) \phi_3(x_3) \rangle_0$ does not vanish. (For the φ^3 model of Sec. 1 we would have $\phi_1 = \phi_2 = \phi_3$, $c_1 = c_2 = c_3$). Then, the non-amputated li-function $G_{1i}(x_1 x_2; x_3 x_4)$ has a "shadow pole" [15] for $c = -c_3$, which is cancelled by a singularity of the 1-particle reducible Green function [M2]. On the other hand, according to (11.8), (13.33), (13.34) the conformal partial wave $s_0(\chi)$ of the disconnected Schwinger function

$s_{11}(x_1 - x_3) s_{22}(x_2 - x_4) + s_{12}(x_1 - x_4) s_{21}(x_2 - x_3)$ is 1. That gives

$$\begin{aligned} s(x_1, x_2, x_3, x_4) &= s_{12}(x_1 - x_2) s_{12}(x_3 - x_4) + \\ &+ 2\pi \sum_{\chi_\ell} \frac{p_\ell(c_\ell)}{\sin\pi(\ell + c_\ell)} \operatorname*{Res}_{\chi = \chi_\ell} \left\{ [1 + g(\chi)] \int (dp)\, Q_+^{\tilde\chi}(x_1, x_2; -p)\, w_\chi(p)\, Q_+^{\tilde\chi}(x_3, x_4; p) \right. \end{aligned} \quad (15.29)$$

where $g(\chi)$ is the conformal partial wave of $G_{1i}(x_1, x_2; x_3, x_4)$ and the sum is over all poles of $A_\ell^2(c_+, c_-; c)[1 + g(\chi)]$ in the right half plane c except the "shadow singularity" for $\chi = \tilde\chi_0 = [0, -c_3]$. We note that for a generalized free field (for which $g(\chi) = 0$) the disconnected Green function $s_{11}(x_1 - x_3) s_{21}(x_2 - x_4) + s_{12}(x_1 - x_4) s_{21}(x_2 - x_3)$ is reproduced by the poles of $\Gamma(h - \delta_\chi + c_+)$ coming from the normalization factors in the Q's. It turns out that for interacting fields these poles are cancelled by zeros of $1 + g(\chi; c_i) = \frac{1}{1 - b(\chi, c_i)}$ i.e. by poles of $b(\chi, c_i)$. This is suggested by the analysis of ultraviolet divergences in the skeleton expansion of the right-hand side of the equation:

$$b(\chi; c_i)\, V(x_3 c_3, x_4 c_4; x\chi) = \frac{1}{2} \int dx_1 \int dx_2\, V(x_1, -c_1, x_2, -c_2; x\chi) B(x_1, x_2, x_3, x_4).$$

(cf. (13.33) (13.34)).

Now we are in a position to analyze the implications of the following (special case of) Osterwalder-Schrader positivity condition for the 4-point function [M3]. Consider the space $\mathcal{S}_+ = \mathcal{S}_+(\mathbb{R}^{2h} \times \mathbb{R}^{2h})$ of test functions $f(x_1, x_2)$ of the Schwartz space $\mathcal{S}(\mathbb{R}^{4h})$ which vanish with all their derivatives unless $\sigma_1 > 0$, $\sigma_2 > 0$ $(\sigma_i = (x_i)_{2h})$ and $x_1 \neq x_2$. Then, for any $f \in \mathcal{S}_+$, we have *)

$$\int \cdots \int dx_1 \ldots dx_4 \, \bar{f}(\theta x_2, \theta x_1) \, (x_1 \ldots x_4) f(x_3, x_4) \geq 0 \qquad [\, \theta(\underline{x}, \sigma) = (\underline{x}, -\sigma) \,] . \tag{15.30}$$

Inserting here for $s(x_1, \ldots x_4)$ its expansion (15.29) we see that this positivity condition is satisfied, provided that the inequality (14.32) takes place and

$$\frac{\rho_\ell(c_\ell)}{\sin \pi(\ell + c_\ell)} \operatorname*{Res}_{\chi = \chi_\ell} g(\chi) > 0 \tag{15.31}$$

for all dynamical poles χ_ℓ of $g(\chi)$, and finally that $[1 + g(\chi)] \operatorname{Res} \Gamma(h + c_+ - \delta_\chi) \geq 0$ at the poles of $\Gamma(h + c_+ - \delta_\chi)$.

16. The problem of crossing symmetry. Concluding remarks

16.A. Crossing symmetry and duality

The vacuum expansion (15.27) or (15.29) of the product $\varphi(x_1)\, \varphi(x_2)$ which satisfies the dynamical equations (in the (1,2)-channel) is not symmetric with respect to a permutation of the arguments x_1 and x_2 with any of the arguments $x_3, \ldots, x_n$. We are stuck here with the analogue of a familiar problem of ordinary partial wave analysis: it simplifies the unitary equations but complicates the crossing symmetry condition. Yet, it should be stressed that the conformal expansion (as pointed out in ref. [M2]) solves an infinite set of coupled non-linear (integral) equations, while the problem of crossing symmetry can be reduced to a set of non-coupled (a finite number for each n) linear (integral) equations for the conformal partial waves.

In order to exhibit these symmetry equations we shall introduce another bit of graphical notation.

We shall represent the Clebsch-Gordan kernel V by

$$V(x_1, -c_1 ; x_2, -c_2 ; x, \tilde{\chi}) = \tag{16.1}$$

---

*) In the original paper of Osterwalder and Schrader condition (15.30) is only assumed to hold for test functions $f(x_1, x_2)$ which vanish unless $\sigma_1 < \sigma_2$. The stronger form of the positivity condition used here is a consequence of analysis of Glaser and Mack [G6,M3] .

and will write Eq. (13.42), (15.1) in the form

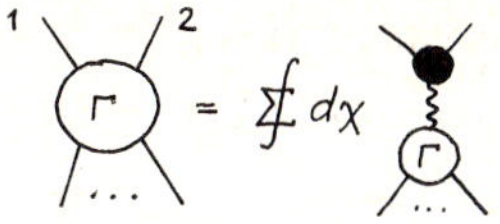

(16.2)

Then the crossing symmetry condition for the special case of the 4-point vertex function assumes the form

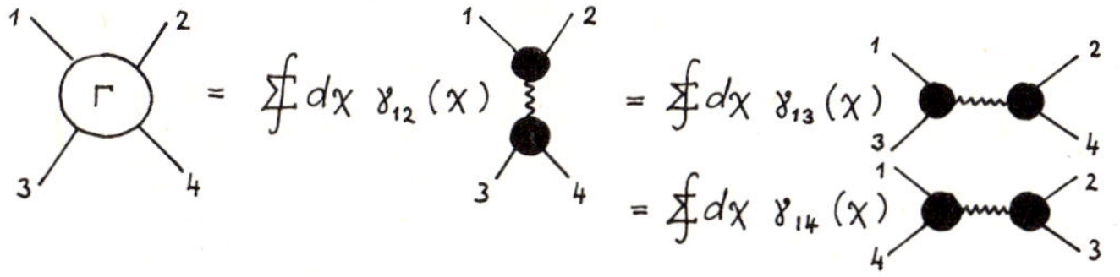

(16.3)

In the case of the $\varphi\varphi^* \to \varphi\varphi^*$ vertex function for the model considered in Sec. 12.D. we would have $\gamma_{12}(\chi) = \gamma_{14}(\chi)$. In the case of the φ^3-model all γ_{ik} should be the same:

$$\gamma_{12}(\chi) = \gamma_{13}(\chi) = \gamma_{14}(\chi) = \gamma(\chi) \quad \text{for} \quad \mathcal{L}_I = -\frac{1}{3!} g :\varphi^3: \tag{16.4}$$

In order to find the crossing symmetry equation for $\gamma(\chi)$ we expand the kernel $F_\chi\,(x_1x_3;x_2x_4)$ (14.3) in conformal partial waves:

$$F_\chi\,(x_1,x_3\,;x_2,x_4) = \sum\!\!\!\!\!\!\int d\chi'\, C(\chi,\chi')\, F_{\chi'}\,(x_1,x_2\,;x_3,x_4) \tag{16.5a}$$

or graphically,

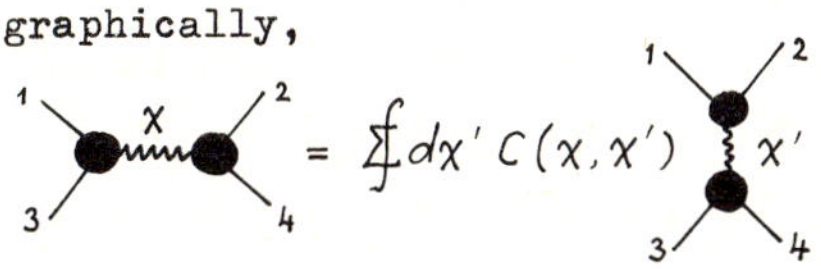

(16.5b)

An explicit expression for the crossing kernel $C(\chi,\chi')$ can be obtained from the following consequence of the orthonormality relation (11.9):

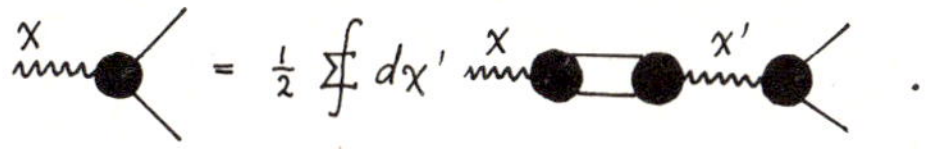

(16.6)

Multiplying both sides of (16.5) by $V(x_1c_1,x_2c_2;x\,\chi'')$ and integrating over x_1 and x_2 we obtain (after replacing χ'' by χ')

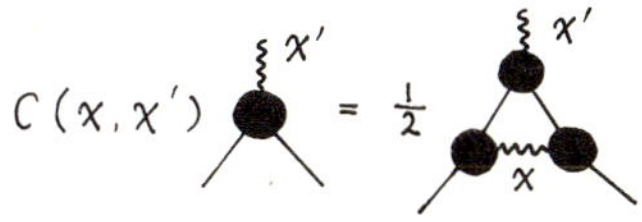

(16.7)

From the involutive property of the crossing operation we deduce that

$$\sum\!\!\!\!\!\!\int d\chi'' \, C(\chi,\chi'')\, C(\chi'',\chi') = \delta(\chi,\chi') \tag{16.8}$$

Inserting (16.5) into (16.3) in the symmetric case (16.4) we obtain the following linear integral equation for $\gamma(\chi)$:

$$\gamma(\chi) = \sum\!\!\!\!\!\!\int d\chi' \, C(\chi',\chi)\, \gamma(\chi') \tag{16.9}$$

We could have alternatively formulated the crossing symmetry condition as a duality property for the discrete vacuum expansion (15.27) (or (15.29)). To do that we first need to continue both sides to Minkowski space arguments with space like separations (since the inequalities (15.3) for different channels contradict each other). That form of crossing symmetry makes obvious its relation to the local commutativity of the underlying fields. For further discussion of this duality property and its extension see [M5'] .

The difficulty in treating the duality relation of type (16.3) comes from the fact that an approximation of Γ involving only a finite number of poles in a given channel would not do. The reason is that the poles of the conformal partial wave in the cross channel are reflected in the divergence of the infinite sum over residues in the direct channel.

16.B. A crossing symmetric representation for the 4-point function

If we forget for a moment about the dynamical equations, it is not difficult to write down a crossing symmetric representation for the conformal partial waves. It can be based on the known Mellin-transform representation of conformal invariant Green functions, proposed by Symanzik [S8] (see also [D2]) and Mansouri [M8]). For instance, the general conformal invariant 4-point function, with dimensions restricted solely by Eq. (13.45), can be written in the form

$$\Gamma(x_1,\dots,x_4) = \left(\frac{4}{x_{13}^2 x_{24}^2}\right)^{\delta} \left(\frac{16}{x_{12}^2 x_{23}^2 x_{14}^2 x_{34}^2}\right)^{\frac{1}{2}(h-\delta)} \left(\frac{2}{x_{12}^2}\right)^{-c_{12}} \left(\frac{2}{x_{34}^2}\right)^{-c_{34}} \left(\frac{2}{x_{13}^2}\right)^{c_{24}} \left(\frac{2}{x_{24}^2}\right)^{c_{13}} \cdot$$
$$\left(\frac{x_{14}^2}{2}\right)^{c_{14}} \left(\frac{x_{23}^2}{2}\right)^{c_{23}} \int_{-\infty}^{\infty} \frac{d\sigma}{2\pi} \int_{-\infty}^{\infty} \frac{d\tau}{2\pi}\, K(\sigma,\tau) \left(\frac{x_{13}^2 x_{24}^2}{x_{12}^2 x_{34}^2}\right)^{i\sigma} \left(\frac{x_{13}^2 x_{24}^2}{x_{14}^2 x_{23}^2}\right)^{i\tau} . \tag{16.10}$$

where $c_{ik} = \frac{1}{2}(c_i + c_k)$ as in (13.46). The right-hand side of (16.10) is independent of the real parameter δ provided that $K(z, w)$ is analytic in a strip domain along the real axes which includes the points $z = \sigma - i\delta$, $w = \tau - i\frac{\delta}{2}$ and that no poles arising from the x-dependent factors prevent us from shifting the integration path. In the φ^3 model under consideration we have

$$c_{ik} = c_\varphi \tag{16.11}$$

The representation (5.10) can be made manifestly crossing symmetric by setting $\delta = \frac{1}{3}h - \frac{4}{3}c_\varphi$ and

$$K(\sigma, \tau) = f(\sigma, \tau, -\sigma-\tau) \tag{16.12a}$$

where

$$f(\alpha, \beta, \gamma) = f(\beta, \alpha, \gamma) = f(\alpha, \gamma, \beta) . \tag{16.12b}$$

Inserting then (16.10) into Eqs. (13.33) (13.34) defining $\gamma(\chi)$ we obtain a conformal partial wave (depending on an arbitrary symmetric function $f(\alpha, \beta, \gamma)$) which satisfies automatically the crossing symmetry equation (16.9)*). The difficulty now is to construct an f consistent with the pole structure of $\gamma(\chi)$ implied by the dynamical equations. This problem is not yet solved.

16.C. Summary and discussion

Our aim in this Chapter has been to construct a conformal invariant quantum field theory satisfying

(a) the dynamical equations (12.11) - (12.20) (in a given channel),

(b) Wightman (or Osterwalder-Schrader) positivity, and

(c) crossing symmetry.

We were able to solve (a) and to incorporate some consequences of (b) by using the vacuum operator product expansion, which can be written in the form

$$\varphi_2(x_2)\varphi_1(x_1)|0\rangle = s(x_1 - x_2)|0\rangle + \sum_{\chi_\ell} C(\chi_\ell) \int dx\, \tilde{Q}_+^{\chi_\ell}(x_1, x_2; x)\, O_{\chi_\ell}(x)|0\rangle , \tag{16.3}$$

where we have set (for Minkowski space coordinates with space-like x_{12})

---

*) Using the integration formula of ref. [S8] we obtain:

$$\gamma(\chi) = \frac{N_\ell(c_{12}, c_-^{12}; c)(2\pi)^{2h}}{2N_\ell(-c_{34}, -c_-^{34}; c)} \int_{-i\infty}^{i\infty} \frac{d\alpha}{2\pi} \int_{-i\infty}^{i\infty} \frac{d\beta}{2\pi} \int_{-i\infty}^{i\infty} \frac{d\sigma}{2\pi} \int_{-i\infty}^{i\infty} \frac{d\tau}{2\pi} \sum_{m=0}^{\ell} \frac{(-)^m \binom{\ell}{m} \Gamma(-\alpha-\beta)\Gamma(\alpha)\Gamma\left(-\alpha-\beta-\sigma-\tau-\frac{h-c+\ell}{2}+\delta+c_{14}\right)\Gamma\left(\alpha+\sigma+\frac{h-\delta}{2}+c_-^{12}\right)}{\Gamma\left(\frac{h+c+\ell}{2}+c_-^{12}\right)\Gamma\left(h-\frac{c+\delta-\ell}{2}+\sigma\right)\Gamma\left(\delta+c_{24}-\sigma-\tau\right)\Gamma\left(\frac{h-\delta}{2}-c_{14}+\tau\right)}$$

$$\cdot \frac{\Gamma\left(\beta+\frac{h-c+\ell}{2}-c_-^{12}\right)\Gamma\left(\beta+\tau+\frac{h-\delta}{2}-c_{14}\right)\Gamma\left(\frac{\delta+\ell-c}{2}-\sigma-\alpha\right)\Gamma\left(\frac{c+\delta+\ell}{2}+c_{13}-\tau-\beta-m\right)\Gamma\left(\frac{h-\delta}{2}-c_{14}+\tau+\beta+m\right)\Gamma\left(\frac{c+\ell+\delta}{2}-\sigma-\alpha\right)}{\Gamma\left(h-\frac{c+\delta-\ell}{2}-c_{13}+\tau+\beta\right)\Gamma\left(h-\frac{c+\delta-\ell}{2}+\sigma+\alpha-m\right)\Gamma\left(\frac{c-\ell+\delta}{2}+m-\alpha-\sigma\right)\Gamma\left(\delta+c_{13}-\alpha-\beta-\sigma-\tau\right)} K(-i\sigma, -i\tau)$$

$$
\cdot \{\Gamma(h-\delta_\chi - c_+)\Gamma(h-\delta_{\tilde\chi}-c_+)\Gamma(h-\delta_\chi - c_-)\Gamma(h-\delta_\chi + c_-)\Gamma(h-\delta_{\tilde\chi}-c_-)\Gamma(h-\delta_{\tilde\chi}+c_-)\}^{1/2} \tilde{Q}_+^{\tilde\chi}(x_1, x_2; x)
$$

$$
= \left(\frac{2}{x_{12}^2}\right)^{h-\delta_{\tilde\chi}+c_+} \mathcal{D}_\ell\left(\delta_{\tilde\chi}+c_-, 3\nabla_1; \delta_{\tilde\chi}-c_-, 3\nabla_2\right) \int (dp)\, \theta_+(p) \left(\frac{-x_{12}^2}{p^2}\right)^{\frac{1}{2}(\ell+c)} \cdot \tag{16.14}
$$

$$
\cdot \int_0^1 du\, [u(1-u)]^{\frac{1}{2}h-1} \left(\frac{u}{1-u}\right)^{c_-} e^{ip(ux_{12}+x_2-x)} J_{\ell+c}\left(\sqrt{-u(1-u)x_{12}^2 p^2}\right)
$$

and

$$
C(\chi_\ell) = \left\{ \frac{2\pi \rho_\ell(c_\ell)}{\sin\pi(\ell+c_\ell)} \operatorname*{Res}_{\chi=\chi_\ell} \left[\Gamma(h-\delta_\chi + c_+)\Gamma(h-\delta_{\tilde\chi}+c_+)\,(1+g(\chi))\right] \right\}^{1/2} \tag{16.15}
$$

$$
\chi = [\ell, c]\,,\ \delta_\chi = \tfrac{1}{2}(h+c-\ell)\,,\ \chi_\ell = [\ell, c_\ell]\,,\ c_\pm = \tfrac{1}{2}(c_1 \pm c_2)\ [\text{for } \varphi_1 = \varphi_2 = \varphi,\ c_- = 0,\ c_+ = c_\varphi]
$$

$O_{\chi_\ell}(x)$ are local (hermitian) tensor fields, whose two-point functions are given by the Fourier transform of (15.13):

$$
\langle O_{\chi_\ell}(x) \otimes O_{\chi_{\ell'}}(x') \rangle_0 = w_{\chi_\ell}(x-x')\, \delta_{\ell\ell'}\, \delta_{c_\ell c_{\ell'}} \tag{16.16}
$$

In the simplest φ^3 model the sum in (16.13) is over even values of ℓ only and the first dynamical pole of $g(\chi)$ for $\ell = 2$ comes from the stress-energy tensor. The positivity condition implies the inequality $c_\ell \geq h+\ell-2$ (14.32) and the reality of the coefficients $C(\chi_\ell)$ (16.15). The sum in (16.13) is over all poles of the expression in square brackets in the right-hand side of (16.15) with positive c_ℓ, except for the scalar shadow pole ($c_0 = -c_\varphi$ in the φ^3 model) which is omitted. We notice that the above construction automatically insures the positivity of the energy spectrum (cf. [L3]).

The crossing symmetry condition implies a set of uncoupled linear integral equations for the conformal partial waves. The simplest of these equations--for the partial wave $\gamma(\chi)$ of the 1PI 4-point function $\Gamma(x_1 \ldots x_4)$-- is given by (16.9) (16.7). It is satisfied by the general crossing symmetric conformal invariant 4-point function (16.10-16.12). However, the problem of displaying simultaneously the pole structure, implied by the dynamical equations, and the permutation symmetry, reflecting the local commutativity of the underlying fields, is not solved. We conjecture that in carrying out a construction which takes into account all three requirements (a)(b)(c) one should be able to discard the φ^3 model as inconsistent. The difficulty of this constructive problem should justify further study of simple soluble models [D1, S4] from the point of view of global operator product expansions presented here.

APPENDIX D

Proof of lemma 10.3.

Let us abbreviate

$$A(m,z) = \int_{U_z} du\, \xi^{\ell}(mu)\, D^j(u^{-1}) \tag{D.1}$$

Let $\hat{z} = (1000)$ and $m(z) \in M$ such that $\check{z} \equiv z/|z| = m(z)\hat{z}$.
From definition (D.1) we deduce the covariance property

$$A(m,z) = D^j(m(z))\, A\left(m(z)^{-1} m\, m(z), \hat{z}\right) D^j(m(z))^{-1}$$

Because of definition (10.18') of $\hat{t}^{js}(x)$, the r.h.s. of the equation of lemma 10.3. has the same covariance property. It suffices then to check it for $z = \hat{z}$.

Let us introduce a canonical basis $\{v^j_{a\nu}\}$ in $\mathcal{V}^j$ etc. Here and later on, letters a,b,... label UIR's of the stability group $U \simeq$ Spin (2h-1) of $\hat{z}$. Matrix-elements $D^j_{a\mu\, b\nu}(m) \equiv (v^j_{a\mu}, D^j(m)\, v^j_{b\nu})$. In particular,

$$D^j_{a\mu\, b\nu}(u) = \delta_{ab}\, d^a_{\mu\nu}(u) \qquad u \in U \tag{D.2}$$

are matrix-elements of UIR's of U and independent of j.
In the following we shall omit indices μ, ν.

In this basis $\pi(ja)_{bc} = \delta_{ab}\delta_{ac}$ for $a,b,c \subset j$. Therefore

$\hat{t}^{ja}_{bc} = \delta_{ab}\delta_{ac}$ for $a \subset \ell,\ a \subset j$. We must then show that

$$\int_U du \sum_{c\subset\ell} D^{\ell}_{cc}(mu)\, D^j_{ab}(u^{-1}) = \sum_{c\subset\ell,j} d(c)^{-1}\, \delta_{ca}\, \delta_{cb}\, D^{\ell}_{cc}(m)$$

But

$$\begin{aligned} \text{l.h.s.} &= \int du \sum_{c\subset\ell} D^{\ell}_{cc}(m)\, d^c(u)\, d^a(u^{-1})\, \delta_{ab} \\ &= \sum_{c\subset\ell} d(c)^{-1}\, \delta_{ac}\, \delta_{ab}\, D^{\ell}_{cc}(m) = \text{r.h.s.} \end{aligned}$$

because of Schur orthogonality relations for representation functions (a, b $\subset$ j).

APPENDIX E

A summation formula involving ratios of Γ-functions

Eqs. (11.16) and (C.19) can be derived from the following known formula for the value of the hypergeometric function $F = {}_2F_1$ at the point $x = 1$:

$$F(a,b;c;1) = \sum_{m=0}^{\infty} \frac{\Gamma(a+m)\Gamma(b+m)\Gamma(c)}{\Gamma(a)\Gamma(b)\Gamma(c+m)\,m!} = \frac{\Gamma(c)\Gamma(c-a-b)}{\Gamma(c-a)\Gamma(c-b)} \tag{E.1}$$

(see e.g. [G7] Eq. 9.122.1).

In order to reduce Eq. (11.16) to the form (E.1) we set $k-j=m$, $\ell-j=n$, and continue to non-integer n, writing it in the form

$$\sum_{m=0}^{\infty} \frac{\Gamma(m-n)\Gamma(\alpha+j+m)}{\Gamma(-n)\,m!\,\Gamma(\beta+j+m)} = \frac{\Gamma(\beta-\alpha+n)\Gamma(\alpha+j)}{\Gamma(\beta+j+n)\Gamma(\beta-\alpha)} \tag{E.2}$$

Here, we have used the identity

$$(-1)^m \frac{\Gamma(n+1)}{\Gamma(n-m+1)} = \frac{\Gamma(m-n)}{\Gamma(-n)} \tag{E.3}$$

Eq. (C.19) is established in a similar way.

There exists also a direct elementary proof of Eqs. (11.16) and (C.19) which exploits their similarity to the Newton binomial formula.

In the above notation, Eq. (11.16) assumes the form

$$f_n(a,b) = \sum_{m=0}^{n} (-1)^m \binom{n}{m} (a)_m (b+m)_{n-m} = (b-a)_n \tag{E.4}$$

where $a = \alpha+j$, $b = \beta+j$ $(n = \ell-j)$ and

$$(x)_k = \frac{\Gamma(x+k)}{\Gamma(x)} = x(x+1)\dots(x+k-1) \tag{E.5}$$

is the finite-difference counter part of the power x^k. In order to prove (E.4), we evaluate the finite difference $f_n(a,b) - f_n(a,b-1)$ using

$$(x)_k - (x-1)_k = [x+k-1-(x-1)]\,(x)_{k-1} = k(x)_{k-1}\,. \qquad \text{(E.6)}$$

Thus, we find the recurrence relation

$$f_n(a,b) - f_n(a,b-1) = nf_{n-1}(a,b) \qquad \text{(E.7)}$$

with the initial condition

$$f_1(a,b) = b-a\,. \qquad \text{(E.8)}$$

In order to fix $f_n(a,b)$ uniquely we have to evaluate it for a particular value of b. For $b = a$ we have

$$f_n(a,a) = (a)_n \sum_{m=0}^{n} (-1)^m \binom{n}{m} = (a)_n\,(1-1)^n = 0 \qquad \text{(E.9)}$$

It is easily seen that the only polynomial solution of (E.7-9) is given by the right-hand side of (E.4).

To reduce (C.19) to (E.4) we multiply both sides by $(-1)^\nu$ and obtain (using (E.3)):

$$\sum_{m=0}^{\nu} (-1)^m \binom{\nu}{m} (-s)_m\, (2-h-\ell-\nu+m)_{\nu-m} = (2-h-\ell-\nu+s)_\nu\,, \qquad \text{(E.10)}$$

which is (E.4) with $a = -s$, $b = 2-h-l-\nu$, $n = \nu$.

APPENDIX F

Partial Fourier transform of $V(x_1,x_2,x_3)$ and related formulas

F.1 Fourier transform in x_3

To evaluate the Fourier transform of V in the third argument, it is convenient to use the representation (11.4b) involving the differential operator D_ℓ -- (11.5).

The calculation of the Fourier transform of V is reduced to the application of the following known relations:

$$\frac{\Gamma(a)\Gamma(b)}{(2\pi)^h}\int\left(\frac{2}{x_{13}^2}\right)^a\left(\frac{2}{x_{23}^2}\right)^b e^{-ipx_3}dx_3 =$$

$$=(2\pi)^{-h}\int_0^\infty\frac{d\alpha}{\alpha}\int_0^\infty\frac{d\beta}{\beta}\alpha^a\beta^b\int dx_3\exp\left\{-\alpha\frac{x_{13}^2}{2}-\beta\frac{x_{23}^2}{2}-ipx_3\right\}= \tag{F.1}$$

$$=\int_0^1 du\, u^{a-1}(1-u)^{b-1}e^{-ip[ux_1+(1-u)x_2]}\int_0^\infty d\lambda\,\lambda^{a+b-h-1}\exp\left\{-\frac{1}{2}\left[\lambda u(1-u)x_{12}^2+\frac{p^2}{\lambda}\right]\right\},$$

$$\int_0^\infty d\lambda\,\lambda^{c-1}\exp\left\{-\frac{1}{2}\left(\lambda\alpha^2+\frac{\beta^2}{\lambda}\right)\right\}=2\left(\frac{\beta}{\alpha}\right)^c K_c(\alpha\beta) \tag{F.2}$$

(see [G7] Eq. 3.471.9). The result is

$$V_+^\chi(x_1,x_2;p,z)=\int V(x_1,c_1;x_2,c_2;x_3,\chi,z)\,e^{-ipx_3}dx_3=$$

$$=2A_\ell\left(\frac{2}{x_{12}^2}\right)^{h-\delta_\chi+c_+}D_\ell(\delta_\chi+c_-,z\nabla_1;\delta_\chi-c_-,z\nabla_2)\left(\frac{x_{12}^2}{p^2}\right)^{\frac{1}{2}(\ell-c)}\times \tag{F.3}$$

$$\times\int_0^1 du\,[u(1-u)]^{\frac{1}{2}h-1}\left(\frac{u}{1-u}\right)^{c_-}e^{-ip(ux_{12}+x_2)}K_{\ell-c}\left(\sqrt{u(1-u)x_{12}^2p^2}\right),$$

where we have used the equation $h-2\delta_\chi=\ell-c$ and have set

$$A_\ell=A_\ell(c_+,c_-;c)=\frac{N_\ell(c_+,c_-,c)}{\Gamma(\delta_\chi+\ell+c_-)\Gamma(\delta_\chi+\ell-c_-)}=A_\ell(c_+,c_-;-c); \tag{F.4}$$

$$A_\ell^\pm=A_\ell(\pm c_+,\pm c_-,c)=\frac{N_\ell(\pm c_+,\pm c_-,c)}{\Gamma(\delta_\chi+\ell+c_-)\Gamma(\delta_\chi+\ell-c_-)}=A_\ell(\pm c_+,\pm c_-,-c)$$

For $x_{12}\to 0$ we can evaluate the main term in (F.1) exactly. The result depends on the sign of Re$c-\ell$. If V corresponds to a physical 3-point function then the Wightman positivity condition (14.32) implies that for $h \geqslant 2$, $c-\ell$ is non-negative. In this case $2K_{\ell-c}(z)\sim\Gamma(c-\ell)\left(\frac{2}{z}\right)^{c-\ell}$ for $z\to 0$ and the small distance behavior of V^χ is given by

$$V_+^\chi(x_1,x_2;p,\zeta)\underset{x_{12}\to 0}{\approx} A_\ell\left(\frac{2}{x_{12}^2}\right)^{h-\delta_\chi+c_+}\mathcal{D}_\ell(\delta_\chi+c_-,\zeta\nabla_1;\delta_\chi-c_-,\zeta\nabla_2)B(h-\delta_\chi+c_-,h-\delta_\chi-c_-)e^{ipx_2}\left(\frac{x_{12}^2}{2}\right)^{\ell-c}$$

$$\approx\frac{\Gamma(c)N_\ell(c_+,c_-,c)}{\Gamma(c-\ell)\Gamma(h-c+\ell)}(h+c-1)_\ell\, e^{ipx_2}\left(\frac{2}{x_{12}^2}\right)^{h-\delta_\chi+c_+}(\zeta x_{12})^\ell \tag{F.5}$$

F.2. Derivation of Eq. (13.36) for the conformal partial wave

We shall first derive Eq. (13.36) for $\gamma(\chi)$ for

$$\mathrm{Re}\, c\geqslant\ell \tag{F.6}$$

and then proceed by analytic continuation.

We start with Eq. (13.33) for $n=4$, and after insertion of (13.34) integrate both sides with respect to x. The result is given by (F.5) with $p=0$ and $c_+\to -c_{34}$, $c_-\to -c_-$ on the left hand side. Noting that (because of (13.35))

$$\frac{N_\ell(c_{12},c_-;c)}{N_\ell(-c_{34},-c_-;c)}=\left\{\frac{\Gamma(h-\delta_\chi+c_{12})\Gamma(h-\delta_{\tilde\chi}+c_{12})\Gamma(h-\delta_\chi+c_{34})\Gamma(h-\delta_{\tilde\chi}+c_{34})}{\Gamma(h-\delta_\chi-c_{12})\Gamma(h-\delta_{\tilde\chi}-c_{12})\Gamma(h-\delta_\chi-c_{34})\Gamma(h-\delta_{\tilde\chi}-c_{34})}\right\}^{1/2}\equiv b_\ell(c_{12},c_{34};c) \tag{F.7}$$

$$\left[c_{ik}=\tfrac{1}{2}(c_i+c_k)\right]$$

we obtain

$$\gamma(\chi)(x_{34}\zeta)^\ell=\frac{1}{2}b_\ell(c_{12},c_{34};c)\left(\frac{x_{34}^2}{2}\right)^{h-\delta_{\tilde\chi}-c_{34}}\int dx_1\int dx_2\left(\frac{2}{x_{12}^2}\right)^{h-\delta_{\tilde\chi}+c_{12}}(\zeta x_{12})^\ell\,\Gamma(x_1x_2x_3x_4) \tag{F.8}$$

Finally, we apply to both sides of (F.8) the operator

$$\frac{1}{\ell!(h-1)_\ell}\left(\frac{x_{34}D}{x_{34}^2}\right)^\ell$$

where D is the interior differentiation (A.47) (A.49) on the complex light-cone, and use (A.13) ($n=2h$)

$$\frac{1}{\ell!(h-1)_\ell}(\eta D)^\ell(\zeta\zeta)^\ell=H_\ell(\eta,\zeta)\equiv(\eta\zeta)^\ell F\left(-\frac{\ell}{2},\frac{1-\ell}{2};2-h-\ell;\frac{\eta^2\zeta^2}{(\eta\zeta)^2}\right)$$

$$=\frac{\ell!}{(h-1)_\ell}\left(\frac{1}{4}\eta^2\zeta^2\right)^{\ell/2}C_\ell^{h-1}\left(\frac{\eta\zeta}{\sqrt{\eta^2\zeta^2}}\right). \tag{F.9}$$

Noting the normalization condition

$$C_\ell^{h-1}(1)=\frac{(2h-2)_\ell}{\ell!} \tag{F.10}$$

for the Gegenbauer polynomials we end up with Eq. (13.36).

F.3. Splitting of $V^{\chi}(x_1,x_2;\ p)$ into two Q functions

Using the known relation (see, e.g. [G7] , Eq. 8.485)

$$2K_\nu(z) = -\frac{\pi}{\sin\pi\nu}\left[I_\nu(z) - I_{-\nu}(z)\right]$$

between modified Bessel functions, we obtain a splitting of V^{χ} into two Q-functions, which have the properties (i) - (iii) of Sec. 15.A. In order to prove Eq. (15.3), (15.4), we will establish the following relation

$$\left(\frac{2}{x_{12}^2}\right)^{\frac{1}{2}(\ell+c)}\left\{\frac{1}{\ell!\,(h-1)_\ell}G_\chi(p;3\,.D')D_\ell(\delta_{\tilde\chi}+c_-,3'\nabla_1;\delta_{\tilde\chi}-c_-,3'\nabla_2)\right\}\left(\frac{x_{12}^2}{p^2}\right)^{\frac{1}{2}(\ell+c)}\cdot$$

$$\cdot\int_{-1}^{1}dt\,\rho_{c_-}(t)\,e^{-ipx_t}I_{\ell+c}\left(\tfrac{1}{2}\sqrt{(1-t^2)x_{12}^2p^2}\right) = \left(\frac{2}{x_{12}^2}\right)^{\frac{1}{2}(\ell-c)}D_\ell(\delta_\chi+c_-,3\nabla_1;\delta_\chi-c_-,3\nabla_2)$$

$$\cdot\left(\frac{p^2}{x_{12}^2}\right)^{\frac{1}{2}(c-\ell)}\int_{-1}^{1}dt\,\rho_{c_-}(t)\,e^{-ipx_t}I_{c-\ell}\left(\tfrac{1}{2}\sqrt{(1-t^2)x_{12}^2p^2}\right) \tag{F.11}$$

where t is related to the integration variable u in (F.3) and (15.4) by $t = 2u-1$, and

$$\rho_{c_-}(t) = \frac{1}{2}\left(\frac{1+t}{2}\right)^{\frac{1}{2}h-1+c_-}\left(\frac{1-t}{2}\right)^{\frac{1}{2}h-1-c_-}, \tag{F.12}$$

$$x_t = \frac{1+t}{2}x_1 + \frac{1-t}{2}x_2 = \frac{1}{2}(x_1+x_2) - \frac{1}{2}t\,x_{12} \tag{F.13}$$

If we assume that (F.11) is true, then multiplying both sides by $A_\ell(c_+,c_-;c)\left(\frac{2}{x_{12}^2}\right)^{c_++\frac{h}{2}}$ and using (F.4) we obtain the counterpart of (15.3) for $V_-^{\tilde\chi}(x_1,x_2;\ -p)$ replaced by $V_+^{\chi}(x_1,x_2;p)$.

The proof of (F.11) is rather tricky, since the equality does not hold for the (t-) integrand. We shall only verify it for the leading terms in both sides for $x_{12} \to 0$. The validity of (F.11) for arbitrary x_1 and x_2 would then follow from the covariance of Q_+^{χ} under the semigroup S defined in ref. [L3] (S consists of those transformations of $O^{\uparrow}(2h+1,1)$ which leave the sign of the Euclidean time component x_{2h} invariant).

To find the small x_{12} behavior of each side of (F.11) we use the power series expansion

$$I_\nu(z) = \left(\frac{z}{2}\right)^\nu \sum_{k=0}^{\infty} \frac{1}{k!\,\Gamma(\nu+k+1)} \left(\frac{z^2}{4}\right)^k \tag{F.14}$$

of the Bessel function and the relations

$$\frac{1}{2}\int_{-1}^{1} dt \left(\frac{1+t}{2}\right)^{a-1} \left(\frac{1-t}{2}\right)^{b-1} = B(a,b) = \frac{\Gamma(a)\Gamma(b)}{\Gamma(a+b)} \tag{F.15}$$

$$\frac{1}{\Gamma(c+\ell+1)} D_\ell\left(\delta_{\tilde{\chi}}+c_-, 3\nabla_1 ; \delta_{\tilde{\chi}}-c_-, 3\nabla_2\right) \left(\frac{x_{12}^2}{2}\right)^{\ell+c} = (-1)^\ell \frac{(h-c-1)_\ell}{\Gamma(c+1)} \left(\frac{x_{12}^2}{2}\right)^c (3x_{12})^\ell \tag{F.16}$$

$$\begin{aligned} \frac{1}{(h-1)_\ell\, \ell!} G_\chi(p; 3, D')\,(x_{12}3')^\ell &= G_\chi(p; 3, x_{12}) = \\ &= \frac{(-1)^\ell\, \ell!}{(h-c-1)_\ell} \left(\frac{p^2}{2}\right)^c \left(\frac{2p3\; p\cdot x_{12}}{p^2}\right)^\ell P_\ell^{(c-\ell,\,h-2)}\left(1 - \frac{p^2\; 3\cdot x_{12}}{p\cdot 3\; p\cdot x_{12}}\right) \end{aligned} \tag{F.17}$$

$$\begin{aligned} &D_\ell\left(\delta_\chi+c_-, 3\nabla_1 ; \delta_\chi-c_-, 3\nabla_2\right)\left[\left(\frac{x_{12}^2}{2}\right)^k e^{-ipx_t}\right] = \\ &= \ell!\,(h+c-1)_k\,(ip3)^{\ell-k}\,(-x_{12}3)^k\, e^{-ipx_t}\, P_{\ell-k}^{(\alpha_k^-,\,\alpha_k^+)}(t) + O\left(x_{12}^2\,(x_{12}\cdot 3)^{k-1}\right), \\ &\qquad \alpha_k^\pm = \delta_k + k - 1 \pm c_- , \quad k = 0, 1, \ldots, \ell . \end{aligned} \tag{F.18}$$

Here $P^{(\alpha,\beta)}(t)$ is the Jacobi polynomial; Eq. (F.17) is a consequence of (13.13) (b)) ; in deriving (F.18) we used the following relation between the Jacobi polynomials and the hypergeometric function

$$P_n^{(\alpha,\beta)}(t) = (-1)^n \frac{(\beta+1)_n}{n!} F\left(n+\alpha+\beta+1, -n; \beta+1; \frac{1+t}{2}\right) \tag{F.19}$$

(see Eq. 8.962.1 of ref. [G7]). Using further the integration formula (we caution the reader that the corresponding equation 7.391.3 of ref. [G7] contains an error):

$$\frac{1}{2}\int_{-1}^{1} dt \left(\frac{1-t}{2}\right)^\alpha \left(\frac{1+t}{2}\right)^{\beta+\nu} P_n^{(\alpha,\beta)}(t) = B(\alpha+n+1, \beta+n+1) \quad \text{for } \nu = n \tag{F.20a}$$

$$= 0 \qquad \text{for } \nu = 0, \ldots, n-1 , \tag{F.20b}$$

we obtain the following small x_{12} expression for the right-hand side of (F.11)

$$\text{RHS} \underset{x_{12}\to 0}{\approx} \mathcal{B}\left(\delta_\chi + \ell + c_-, \delta_\chi + \ell - c_-\right) \exp\left\{-ip\,\frac{x_1+x_2}{2}\right\} \cdot$$

$$\cdot \left(\frac{2}{x_{12}^2}\right)^{\frac{1}{2}(\ell - c)} \left(\frac{p^2}{2}\right)^c \sum_{k=0}^{\ell} \binom{\ell}{k} \frac{(h+c-1)_k}{\Gamma(c+k-\ell+1)} \left(\frac{2p\mathfrak{z}\; px_{12}}{p^2}\right)^{\ell-k} (-x_{12}\mathfrak{z})^k . \tag{F.21}$$

It follows from (F.15-17) and from the identity

$$\ell!\, P_\ell^{(c+\ell,\, h-2)}(\omega) = (c+1-\ell)_\ell\, F\left(h+c-1, -\ell\,; c+1-\ell\,; \frac{1-\omega}{2}\right) ,$$

$$\left(\omega = 1 - \frac{p^2\, \mathfrak{z} x_{12}}{p\mathfrak{z}\; px_{12}}\right) \tag{F.22}$$

(see Eq. 8.962.2 of ref. [G7]) that the left-hand side of (F.11) is also given by (F.21) in the small x_{12} limit. This completes our proof of the representation (15.3).

APPENDIX G

Identities between Q and γ functions for partially equivalent representations

Eq. (15.21) is equivalent to (11.24). To see this note that

$$(?pz)^\nu Q_-^\chi(x_1,x_2;-pz) = [z(\nabla_1+\nabla_2)]^\nu Q_-^\chi(x_1,x_2;-p,z) \qquad (G.1)$$

because $z(\nabla_1+\nabla_2) f(x_{12}) = 0.$

Now it remains to prove

$$[z(\nabla_1+\nabla_2)]^\nu A_\ell(-c_+,-c_-,1-h-\ell-\nu) D_\ell(\delta_{\ell\nu}-c_-, z\nabla_1; \delta_{\ell\nu}+c_-, z\nabla_2) =$$
$$= \operatorname{sign}\left[\left(\tfrac{1-\nu}{2}+c_-\right)_\nu\right] A_{\ell+\nu}(-c_+,-c_-,1-h-\ell) D_{\ell+\nu}(\delta_{\ell\nu}-c_-, z\nabla_1; \delta_{\ell\nu}+c_-, z\nabla_2) \qquad (G.2)$$

where $\delta_{\ell\nu} = \frac{1-\nu}{2}-\ell$ is given in (11.26). But this is in fact what we have proved in the derivation of (11.24) (with $c_\pm \to -c_\pm$) noting

$$\left(\tfrac{1-\nu}{2}-c_-\right)_\nu = (-1)^\nu\left(\tfrac{1-\nu}{2}+c_-\right)_\nu \qquad (G.3)$$

Passing in (13.33) to pseudo-euclidean $p \to p_M$ (cf. 15.A) and using the definition (15.13) for the kernels $Q_r^\chi(p_M;x_3,\dots,x_n)$ and Eq. (15.9′) we obtain

$$Q_r^\chi(p_M, x_3,\dots x_n) = \iint Q_-^\chi(x_1,x_2;p_M)\,\Gamma(x_1,x_2,\dots,x_n)\,dx_1dx_2\,. \qquad (G.4)$$

This equation together with (15.21) leads to (15.22) (in the same way, as we derived (11.24′)).

In the special case n=4 using (13.34) and comparing (15.21) and (15.22) we find

$$\left(\tfrac{1-\nu}{2}+c_-\right)_\nu \gamma(\chi_{\ell\nu}^-) = \left(\tfrac{1-\nu}{2}-c_-\right)_\nu \gamma(\chi_{\ell\nu}^{\prime -}). \qquad (G.5)$$

We note that for $c_-=0$ Eq. (G.2) as well as (15.22) require a modification, since

$$\left(\tfrac{1-\nu}{2}\right)_\nu = 0 \qquad \text{for odd } \nu, \qquad (G.6)$$

and the sign function in the above formulas is not defined (cf. the remark at the end of Sec. 15.B).

REFERENCES

A1 S. Adler, Massless, Euclidean quantum electrodynamics on the five-dimensional unit hypersphere, Phys. Rev. D6, 3445 (1972); Errata, ibid. D7, 3821 (1973).

B1 V. Bargmann, Irreducible unitary representations of the Lorentz group, Ann. of Math. vol. 48 (1947).

B2 V. Bargmann and I.T. Todorov, Spaces of analytic functions on a complex cone as carriers for the symmetric tensor representations of SO(n), (to be published).

B3 I. Bars and F. Gürsey, Operator treatment of the Gel'fand-Naimark basis for SL(2,C), J. Math. Phys. 13, 131 (1972).

B4 J.D. Bjorken and S.D. Drell, Relativistic Quantum Fields (McGraw-Hill, N.Y., 1965) Chapter 19 (see in particular Sec. 19.4).

B5 N.N. Bogolubov, A.A. Logunov and I.T. Toderov, Introduction to Axiomatic Quantum Field Theory (W.A. Benjamin, Reading, Mass. 1975). See in particular Ch. 18.

B6 L. Bonora, S. Ciccariello, G. Sartori, M. Tonin, Conformal Symmetry in Wilson operator product expansions, in: Scale and Conformal Symmetry in Hadron Physics, ed. R. Gatto (Wiley, N.Y. 1973).

C1 W. Casselman (to be published).

C2 S. Coleman, Secret symmetry: an introduction to spontaneous symmetry breakdown and gauge fields. Lectures at the 1973 International Summer School in Physics "Ettore Majorana" (to be published).

D1 G.F. Dell'Antonio, Y. Frishman and D. Zwanziger, Thirring model in terms of currents: solution and light-cone expansion, Phys. Rev. D6, 988 (1972).

D2 M. D'Eramo, L. Peliti and G. Parisi, Theoretic predictions for critical exponents at the λ-point of Bose liquids, Nuovo Cim. Lett. 2, 878 (1971).

D3 J. Dixmier, Représentation intégrables du groupe de De Sitter, Bull. Soc. Math. France, V.89 (1961).

D4 V.Dobrev, V. Petkova, S. Petrova and I. T. Todorov, On the exceptional integer points in the representation space of the pseudo-orthogonal group $O^{\uparrow}(2h+1,1)$, ICTP, Trieste preprint IC/75/1 (1975).

D5 V. Dobrev, G. Mack, V. Petkova, S. Petrova and I. T. Todorov, On the Clebsch-Gordan expansion for the Lorentz group in n dimensions, IAS preprint, Princeton (1975) (submitted to Reports Math. Phys.).

D6 V. Dobrev, V. Petkova, S. Petrova and I. T. Todorov, Dynamical derivation of vacuum operator product expansion in Euclidean conformal quantum field theory, IAS preprint, Princeton (1975), Phys. Rev. D (to be published).

F1 S. Ferrara, A. Grillo and R. Gatto, Tensor representations of conformal algebra and conformally covariant operator product expansions, Ann. Phys. (N. Y.) 76, 161 (1973).

F2 S. Ferrara, R. Gatto and A. Grillo, Positivity restrictions on anomalous dimensions, Phys. Rev. D9, 3567 (1974).

F3 S. Ferrara, A. Grillo and R. Gatto, Conformal algebra in space-time and operator product expansion, Springer Tracts in Modern Physics 67 (1973), Properties of partial wave amplitudes in conformal invariant field theories, Nuovo Cim. A (to be published).

F4 S. Ferrara, R. Gatto, A. F. Grillo and G. Parisi, The shadow operator formalism for conformal algebra. Vacuum expectation values and operator products, Lett. Nuovo Cim. 4, 115 (1972).

F5 S. Ferrara, R. Gatto, A.F. Grillo, G. Parisi: General consequences of conformal algebra, in Scale and Conformal Symmetry in Hadron Physics, ed. R. Gatto (Wiley, N. Y. 1973).

F6 J. Fisher, J. Niederle and R. Rączka, Generalized spherical functions for the noncompact rotation groups, J. Math. Phys. 7, 816 (1966).

F7 E. S. Fradkin, The quantum theory of fields, I, Zh. Exp. Teor. Fis. 29, 121 (1955) (translation JETP, 2, 148 (1956)); Green functions in a quantum field theory and quantum statistics, in: Quantum Field Theory and Hydrodynamics (Trudy FIAN, vol. 29, Moscow, 1965, in Russian) pp. 7-138 (transl. Plenum Publ. Corp. N. Y., 1967).

F8 E. S. Fradkin and M. Ya. Palchik, Coformal invariant solution of quantum field theory equations 1 (in Russian), Inst. Automatics and Electrometry, preprint No 74, Novosibirsk 1974.

G1 A. M. Gavrilik, A. U. Klimyk, Analysis of the representations of the Lorentz and Euclidean groups of n-th order,ITP, Kiev preprint, ITP-75-18E (1975).

G2 I. M. Gel'fand, M. I. Graev and N. YA. Vilenkin, Generalized Functions Vol. 5 (Academic Press, New York, 1966).

G3 I. M. Gel'fand and M. A. Naimark, Unitare Darstellungen der klassischen Gruppen (Akademie Verlag, Berlin,1957).

G4 I. M. Gel'fand and G. E. Shilov, Generalized Functions Vol. 1 (Academic Press, New York, 1964).

G5 I. M. Gel'fand and M. L. Zetlin, Finite dimensional representations of the groups of orthogonal matrices (in Russian), DAN SSSR 71, 1017 (1950).

G6 V. Glaser, On the equivalence of the Euclidean and Wightman formulation of field theory, Commun. Math. Phys. 37, 257 (1974)

G7 I. S. Gradshteyn and I. M. Ryzhik, Tables of Integrals, Series and Products (Academic Press, New York, 1965).

G8 G. Grensing, Representation theory of the universal covering of the Euclidean conformal group and conformal invariant Green's functions, J. Math. Phys. 16, 312 (1975).

G9 A. F. Grillo, Conformal invariance in quantum field theory, Rivista Nuovo Cim. 3, 146 (1973).

G10 F. Gürsey and S. Orfanidis, Conformal invariance and field theory in two dimensions, Phys. Rev. D7, 2414 (1973).

H1 Harish-Chandra, Representations of a semi-simple Lie group on a Banach space I -VI, Trans. Amer. Math. Soc. 75, 185 (1953); ibid. 76, 26 (1954); ibid. 76, 236 (1954); Amer. J. Math. 77, 734 (1955); ibid. 78, 1 and 564 (1956).

H2 Harish-Chandra, Discrete series for semi-simple Lie groups I-II, Acta Math. 113, 241 (1965); ibid. 116, 1 (1966).

H3 T. Hirai, On infinitesimal operators of irreducible representations of the Lorentz group of n-th order, Proc. Japan Acad. v. 38, 83 (1962); On irreducible representations of the Lorentz group of n-th order, Proc. Japan Acad. v. 38, 258 (1962).

H4 T. Hirai, The characters of irreducible representations of the Lorentz group of n-th order, Proc. Japan Acad. v. 41, 526 (1965).

H5 T. Hirai, The Plancherel formula for the Lorentz group of n-th order, Proc. Japan Acad. v. 42, 323 (1965).

H6 M. Hortaçsu, R. Seiler and B. Schroer, Conformal symmetry and reverberations, Phys. Rev. D5, 2518 (1972).

H7 R. Hotta, On realization of the discrete series for semi-simple Lie groups, J. Math. Soc. Japan, v. 23, №2, 384 (1971).

K1 E. Kamke, Differentialgleichungen. Lösungsmethoden und Lösungen I. Gewönliche Differentialgleichungen, (Akademische Verlagsgesellschaft, Leipzig, 1943), (see Eq. 2.260 (19)).

K2 A. Kihlberg and S. Ström, On the unitary irreducible representations of the (1+4) de Sitter group, Arkiv f. Fysik 31, 491 (1965).

K3 A. W. Knapp and E. M. Stein, Intertwining operators for semi-simple groups, Ann. of Math. (2) 93, 489 (1971).

K4 A.W. Knapp and E.M. Stein, Existence of complementary series, Problems in Analysis: Symposium in Honor of Salomon Bochner, Princeton University Press, Princeton, New Jersey, 249 (1970).

K5 K. Koller, The significance of conformal inversion in quantum field theory, Commun. Math. Phys. 40, 15 (1975), see also [G7].

K6 J. Kupsch, W. Rühl, B.C. Yunn, Conformal invariance in quantum fields in two-dimensional space-time, Ann. of Phys. 89, 115 (1975).

K7 J. Kurian, N. Mukunda and E. Sudarshan, Master analytic representations and unified representation theory of certain orthogonal and pseudo-orthogonal groups, Commun. Math. Phys. 8, 204, (1968).

L1 S. Lang, $SL_2(R)$ (Addison-Wesley, Reading, Mass., 1975).

L2 J. Lepowski, Algebraic results on representations of semisimple Lie groups Trans. AMS 176, I (1973)

L3 M. Lüscher and G. Mack, Global conformal invariance in quantum field theory, Commun. Math. Phys. 41, 803 (1975), see esp. Appendix C.

M1 G. Mack, Conformal invariance and short distance behaviour in quantum field theory, in : Lecture Notes in Physics, Vol. 17, Strong Interaction Physics (Springer Verlag, Berlin, 1972) p. 485.

M2 G. Mack, Group theoretical approach to conformal invariant quantum field theory, J. de Physique 34, C1 (Suppl. No. 10) 99 (1973) and in : Renormalization and Invariance in Quantum Field Theory, Ed. by E.R. Caianiello, (Plenum Press, New York, 1974), pp. 123-157.

M3 G. Mack, Osterwalder - Schrader positivity in conformal invariant quantum field theory, in: Lecture notes in physics, Vol. 37, H. Rollnik and K. Dietz (Eds.), Springer Verlag, Heidelberg 1975. For background on the Wightman positivity in the Euclidean framework see [O2], [O1].

M4 G. Mack, On Blattner's formula for the discrete series representations of SO(2n,1) J. Functional Analysis 23, 311 (1976)

M5 G. Mack, Conformal covariant quantum field theory, in Scale and Conformal Symmetry in Hadron Physics, ed. R. Gatto (Wiley, N.Y. 1973).

M5' G. Mack, Duality in QFT, Nucl. Phys. (in press).

M6 G. Mack and K. Symanzik, Currents, stress tensor and generalized unitary in conformal invariant quantum field theory, Commun. Math. Phys. 27, 247 (1972).

M7 G. Mack and I.T. Todorov, Conformal invariant Green functions without ultra-violet divergences, Phys. Rev. D8, 1764 (1973).

M8 F. Mansouri, Dual models with global SU(2,2) symmetry, Phys. Rev. D8, 1159 (1973).

M9 A.A. Migdal, On hadronic interactions at small distances, Phys. Letters 37B, 98 (1971); Conformal invariance and bootstrap, ibid. 37B, 386 (1971).

M10 A.A. Migdal, 4-dimensional soluble models of conformal field theory, preprint, Landau Institute for Theoretical Physics, Chernogolovka (1972).

M11 C.C. Moore, Restrictions of unitary representations to subgroups and ergodic theory: group extension and group cohomology, in: Group Representations in Mathematics and Physics, Battelle Seattle 1969 Rencontres, Ed. by V. Bargmann (Springer Verlag, Berlin, 1970) pp. 1-35

N1 M.A. Naimark, Decomposition of the tensor product of proper Lorentz group representations into irreducible representations. Trudy Moskov. Mat. Obshch., 8,121 (1959); ibid. 10, 181 (1961); (English translation: Am. Math. Soc. Transl., 36, 101 and 189 (1964); DAN SSSR 130, 261 (1960), ibid. 132, 1027 (1960); (Soviet Math. Doklady, 1, 713 (1960))

N2 M.A. Naimark, Decomposition of the tensor product of proper Lorentz group representations into irreducible representations. Part I. Trudy Moskov. Mat. Obshch. 8, 121 (1959). (For transl. see [N1] .)

01 A.I. Oksak and I.T. Todorov, General properties of physical region spinor amplitudes, Reports Math. Phys. 7 , 417 (1975).

02 S. Orfandis, The analytic representations of the conformal group, in: John Hopkins University Workshop on Current Problems in High Energy Particle Theory, p. 109 (John Hopkins Univ. 1974).

03 K. Osterwalder, Euclidean Green's functions and Wightman distributions, in: Lecture Notes in Physics, 25, Constructive Quantum Field Theory, ed. G. Velo and A. Wightman (Springer Verlag, Berlin, 1973), pp. 71-93 and references therein.

04 K. Osterwalder and R. Schrader, Axioms for Euclidean Green's functions, Commun. Math. Phys. 31, 83 (1973); Axioms for Euclidean Green's functions II, ibid. 42, 281 (1975).

05 V. Ottoson, A classification of the unitary irreducible representations of SO (N,1), Commun. Math. Phys. 8, 228 (1968).

P1 G. Parisi and L. Peliti, Calculation of critical indices. Lett. Nuovo Cim. 2, 627 (1971).

P2 A.M. Polyakov, Conformal symmetry of critical fluctuations, Zh ETF Pis. Red. 12, 538 (1970). (English translation JETP Lett. 12, 381 (1970).)

P3 A.M. Polyakov, Non-Hamiltonian approach to the quantum field theory at small distances, ZhETF 66, 23 (1974). (English translation JETP 39, 10 (1974).)

R1 M. Reed and B. Simon, Methods of Modern Mathematical Physics. I. Functional Analysis (Academic Press, New York, 1972).

R2 W. Rühl, Field representations of the conformal group with continuous mass spectrum, Commun. Math. Phys. 30, 287 (1973) (see also the more general discussion: W. Rühl, On conformal invariance of interacting fields, ibid. 34, 149 (1973)).

S1 G. Schiffmann, Intégrales d'entrelacement et fonctions de Whittaker, Bull. Soc. Math. France, v.99, 3 (1971).

S2 W. Schmid, Rice University Studies 56, 99 (1970).

S3 B. Schroer, Bjorken scaling and scale invariant quantum field theories, in: Scale and Conformal Symmetry in Hadron Physics, ed. R. Gatto (Wiley, N.Y. 1973)

S4 B. Schroer, J.A. Swieca and A.H. Völkel, Global operator expansions in conformally invariant relativistic quantum field theory, Berlin preprint, FUB HEP Oct. 74/13.

S5 I. Segal, Causally oriented manifolds and groups, Bull. Amer. Math. Soc. 77, 958 (1971).

S6 K. Symanzik, Green functions method and renormalization of renormalizable quantum theory, in : Lectures in High Energy Physics, ed. B. Jaksic, Zagreb, 485 (1961) (Gordon and Breach, N.Y. , 1965).

S7 K. Symanzik, Euclidean quantum field theory, in: Coral Gables Conference on Fundamental Interactions at High Energy, Ed. T. Gudehus et al. (Gordon and Breach, N.Y., 1969), pp. 19-31.

S8 K. Symanzik, On calculations in conformal invariant field theories, Nuovo Cim. Lett. 3, 734 (1972).

T1 R. Takahashi, Sur les représentations unitaires des groups de Lorentz généralisés, Bull. Soc. Math. France v. 91, 289 (1963).

T2 E. Thieleker, On the quasi-simple irreducible representations of the Lorentz groups, Trans. Amer. Math. Soc. 179, 465 (1973).

T3 E. Thieleker, The unitary representations of the generalized Lorentz group, Trans. Amer. Math. Soc. 199, 327 (1974).

T4 I.T. Todorov, Conformal invariant Euclidean quantum field theory, Acta Phys. Austr. Suppl. 11, 241 (1973). See also [T5] .

T5 I.T. Todorov, Conformal invariant quantum field theory with anomalous dimensions (Cargèse Lecture notes) Ref. TH. 1697-CERN (1973). (This reference contains an extensive bibliography.)

T6 I.T. Todorov and R.P. Zaikov, Spectral representation of the covariant two-point function and infinite -component fields with arbitrary mass spectrum, J. Math. Phys. 10, 2014 (1969).

V1 N.Ya. Vilenkin, Special functions related to the class 1 representations of isometry groups of spaces of constant curvature.

Trudy Moskov. Mat. Obshch. 12, 185 (1963) (in Russian).

W1 B.L. van der Waerden, Die Gruppentheoretische Methode in der Quantenmechanik (Springer, Berlin, 1932).

W2 N. Wallach, Harmonic Analysis on Homogeneous Spaces, (Marcel Dekker, Inc. , New York, 1973).

W3 G. Warner, Harmonic Analysis on Semi-simple Lie Groups I,II (Springer-Verlag, Berlin, 1972).

W4 H. Weyl, The Classical Groups.Their Invariants and Representations (Princeton Univ. Press, Princeton, 1939).

W5 F.L. Williams, Tensor Products of Principal Series Representations, Lecture Notes in Mathematics 358, Springer-Verlag, Berlin 1973.

Z1 R.P. Zaikov, Spectral representation of the conformal covariant two-point function for fields with arbitrary spin, Bulgarian Journal of Phys. 2, 89 (1975), Conformal invariant Euclidean two-point function for tensor fields, JINR, Dubna, preprint E2-8241 (1974) (submitted to Reports Math. Phys.).

Z2 D.P. Zhelobenko, Compact Lie Groups and Their Representations (AMS, Providence, Rhode Island, 1973).

Z3 D.P. Zhelobenko, <u>Harmonic Analysis on Complex Semi Simple Lie Groups</u> (Nauka, Moscow 1974, in Russian).

Z4 G.J. Zuckerman, Some character identities for semi-simple Lie groups. Dissertation, Department of Mathematics, Princeton University (1974).

Z5 G.J. Zuckerman, private communication and to be published.

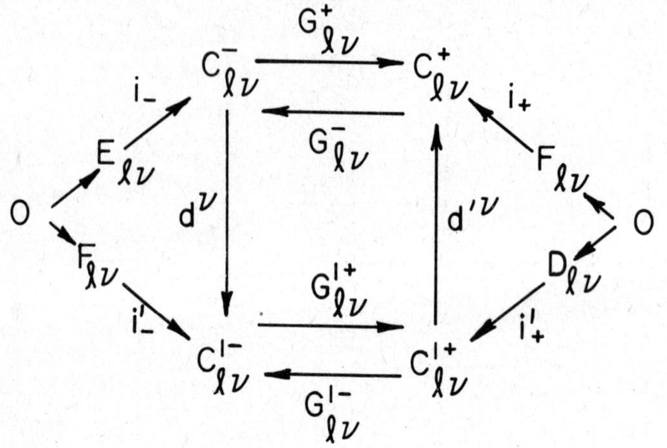

FIG. 1

Figure 1. Quartet diagram of intertwining operators at exceptional points.

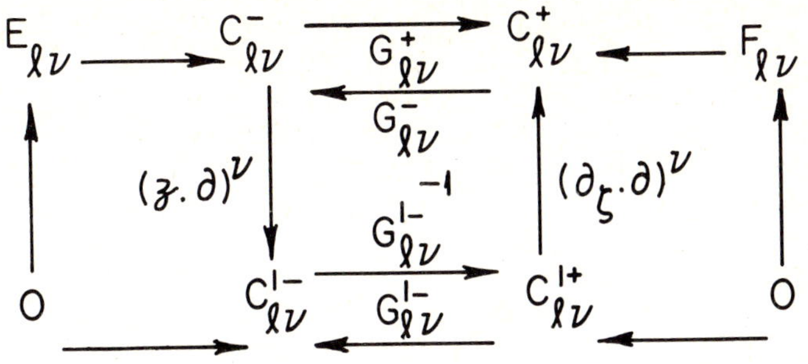

FIG. 2

Figure 2. The quartet diagram for the Lorentz group (h = 1) and $\ell > 0$.

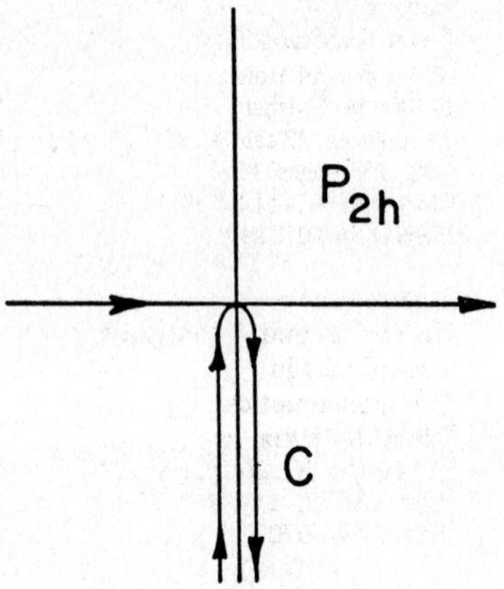

FIG.3

Figure 3. Deformation of the integration path in the complex energy plane (Sec. I5.A). Original path: the real p_{2h}-axis; deformed path: the contour C.

Topics in Applied Physics

Founded by **Helmut K. V. Lotsch**

This book series is devoted to research achievements of current interest. Each volume deals with a different topic under the editorship of a recognized authority in the field. It covers application-oriented aspects of the topic under consideration, the basic physical principles being summarized in a comprehensive introduction.
The contributors to each volume are internationally known experts. The publication periods are comparable with those of scientific journals to keep pace with the rapidly accumulating results.

Springer-Verlag
Berlin
Heidelberg
New York

Volume 1
Dye Lasers
Editor: **F.P. Schäfer**
114 figures. XI, 285 pages. 1973
Cloth DM 77,–; US $ 33.90

Volume 2
Laser Spectroscopy
of Atoms and Molecules
Editor: **H. Walther**
137 figures, 22 tables
XVI, 383 pages. 1976
Cloth DM 97,–; US $ 42.70
ISBN 3-540-07324-8

Volume 3
Numerical and Asymptotic Techniques in Electromagnetics
Editor: **R. Mittra**
112 figures. XI, 260 pages. 1975
Cloth DM 72,–; US $ 31.70
ISBN 3-540-07072-9

Volume 4
Interactions on Metal Surfaces
Editor: **R. Gomer**
112 figures. XI, 310 pages. 1975
Cloth DM 78,–; US $ 34.40
ISBN 3-540-07094-X

Volume 5
Mössbauer Spectroscopy
Editor: **U. Gonser**
96 figures. XVIII, 241 pages. 1975
Cloth DM 70,–; US $ 30.80
ISBN 3-540-07120-2

Volume 6
Picture Processing and Digital Filtering
Editor: **T.S. Huang**
113 figures. XIII, 289 pages. 1975
Cloth DM 79,80; US $ 35.20
ISBN 3-540-07202-0

Volume 7
Integrated Optics
Editor: **T. Tamir**
99 figures. XIII, 315 pages. 1975
Cloth DM 79,80; US $ 35.20
ISBN 3-540-07297-7

Volume 8
Light Scattering in Solids
Editor: **M. Cardona**
111 figures, 3 tables
XIII. 339 pages. 1975
Cloth DM 92,60; US $ 40.80
ISBN 3-540-07354-X

Volume 9
Laser Speckle and Related Phenomena
Editor: **J.C. Dainty**
133 figures. XIII, 286 pages. 1975
Cloth DM 94,80; US $ 41.80
ISBN 3-540-07498-8

Volume 10
Transient Electromagnetic Fields
Editor: **L.B. Felsen**
111 figures. XIII, 274 pages. 1976
Cloth DM 92.60; US $ 40.80
ISBN 3-540-07553-4

Volume 11
Digital Picture Analysis
Editor: **A. Rosenfeld**
114 figures. 47 tables.
XIII, 351 pages. 1976
Cloth DM 72,–; US $ 31.70
ISBN 3-540-07579-8

Volume 12
Turbulence
Editor: **P. Bradshaw**
47 figures, XI, 335 pages. 1976
Cloth DM 97,–; US $ 42.70
ISBN 3-540-07705-7

Volume 13
High-Resolution Laser Spectroscopy
Editor: **K. Shimoda**
132 figures. XIII, 378 pages. 1976
Cloth DM 97,–; US $ 42.70
ISBN 3-540-07719-7

Volume 14
Laser Monitoring of the Atmosphere
Editor: **E.D. Hinkley**
84 figures. XV, 380 pages. 1976
Cloth DM 97,–; US $ 42.70
ISBN 3-540-07743-X

Volume 15
Radiationless Processes in Molecules and Condenses Phases
Editor: **F.K. Fong**
67 figures XIII, 360 pages. 1976
Cloth DM 97,–; US $ 42.70
ISBN 3-540-07830-4

Volume 16
Nonlinear Infrared Generation
Editor: **Y.-R. Shen**
134 figures. XI, 279 pages. 1977
Cloth DM 88,–; US $ 38.80
ISBN 3-540-07945-9

Prices are subject to change without notice

Selected Issues from

Lecture Notes in Mathematics

Lecture Notes in Physics

Vol. 44: R. A. Breuer, Gravitational Perturbation Theory and Synchrotron Radiation. VI, 196 pages. 1975.

Vol. 45: Dynamical Concepts on Scaling Violation and the New Resonances in e^+e^- Annihilation. Edited by B. Humpert. VII, 248 pages. 1976.

Vol. 46: E. J. Flaherty, Hermitian and Kählerian Geometry in Relativity. VIII, 365 pages. 1976.

Vol. 47: Padé Approximants Method and Its Applications to Mechanics. Edited by H. Cabannes. XV, 267 pages. 1976.

Vol. 48: Interplanetary Dust and Zodiacal Light. Proceedings 1975. Edited by H. Elsässer and H. Fechtig. XII, 496 pages. 1976.

Vol. 49: W. G. Harter and C. W. Patterson, A Unitary Calculus for Electronic Orbitals. XII, 144 pages. 1976.

Vol. 50: Group Theoretical Methods in Physics. 4th International Colloquium. Nijmegen 1975. Edited by A. Janner, T. Janssen, and M. Boon. XIII, 629 pages. 1976.

Vol. 51: W. Nörenberg und H. A. Weidenmüller. Introduction to the Theory of Heavy-Ion Collisions. IX, 273 pages. 1976.

Vol. 52: M. Mladjenović, Development of Magnetic β-Ray Spectroscopy. X, 282 pages. 1976.

Vol. 53: D. J. Simms and N. M. J. Woodhouse, Lectures on Geometric Quantization. V, 166 pages. 1976.

Vol. 54: Critical Phenomena. Sitges International School on Statistical Mechanics, June 1976. Edited by J. Brey and R. B. Jones. XI, 383 pages. 1976.

Vol. 55: Nuclear Optical Model Potential. Proceedings 1976. Edited by S. Boffi and G. Passatore. VI, 221 pages. 1976.

Vol. 56: Current Induced Reactions. International Summer Institute, Hamburg 1975. Edited by J. G. Körner, G. Kramer, and D. Schildknecht. V, 553 pages. 1976.

Vol. 57: Physics of Highly Excited States in Solids. Proceedings 1975. Edited by M. Ueta and Y. Nishina. IX, 391 pages. 1976.

Vol. 58: Computing Methods in Applied Sciences. Proceedings 1975. Edited by R. Glowinski and J. L. Lions. VIII, 593 pages. 1976.

Vol. 59: Proceedings of the Fifth International Conference on Numerical Methods in Fluid Dynamics. 1976. Edited by A. I. van de Vooren and P. J. Zandbergen. VII, 459 pages. 1976.

Vol. 60: C. Gruber, A. Hintermann, and D. Merlini, Group Analysis of Classical Lattice Systems. XIV, 326 pages. 1977.

Vol. 61: International School on Electro and Photonuclear Reactions I. Edited by C. Schaerf. VIII, 650 pages. 1977.

Vol. 62: International School on Electro and Photonuclear Reactions II. Edited by C. Schaerf. VIII, 301 pages. 1977.

Vol. 63: V. K. Dobrev et al., Harmonic Analysis on the n-Dimensional Lorentz Group and Its Application to Conformal Quantum Field Theory. X, 280 pages. 1977.

This series aims to report new developments in physical research and teaching – quickly, informally and at a high level. The type of material considered for publication includes:

1. Preliminary drafts of original papers and monographs
2. Lectures on a new field, or presenting a new angle on a classical field
3. Seminar work-outs
4. Reports of meetings, provided they are
   a) of exceptional interest and
   b) devoted to a single topic.

Texts which are out of print but still in demand may also be considered if they fall within these categories.

The timeliness of a manuscript is more important than its form, which may be unfinished or tentative. Thus, in some instances, proofs may be merely outlined and results presented which have been or will later be published elsewhere. If possible, a subject index should be included. Publication of Lecture Notes is intended as a service to the international physical community, in that a commercial publisher, Springer-Verlag, can offer a wider distribution to documents which would otherwise have a restricted readership. Once published and copyrighted, they can be documented in the scientific literature.

Manuscripts

Manuscripts should comprise not less than 100 pages.
They are reproduced by a photographic process and therefore must be typed with extreme care. Symbols not on the typewriter should be inserted by hand in indelible black ink. Corrections to the typescript should be made by pasting the amended text over the old one, or by obliterating errors with white correcting fluid. Authors receive 50 free copies and are free to use the material in other publications. The typescript is reduced slightly in size during reproduction; best results will not be obtained unless the text on any one page is kept within the overall limit of 18 x 26.5 cm (7 x 10½ inches). The publishers will be pleased to supply on request special stationery with the typing area outlined.

Manuscripts in English, German or French should be sent to Prof. Dr. W. Beiglböck, Institut für Angewandte Mathematik, Im Neuenheimer Feld 5, 6900 Heidelberg/Germany, or directly to Springer-Verlag Heidelberg.

Springer-Verlag, Heidelberger Platz 3, D-1000 Berlin 33
Springer-Verlag, Neuenheimer Landstraße 28–30, D-6900 Heidelberg 1
Springer-Verlag, 175 Fifth Avenue, New York, NY 10010/USA

ISBN 3-540-08150-X
ISBN 0-387-08150-X